ANOMALOUS

Exploring UFOs, Non-Human-Intelligence, and Related Phenomena.

By
D.T. Weiss

Paperback: 978-1-964898-64-3

eBook: 978-1-964898-77-3

Published by: Pine Book Writing

www.PineBookWriting.com

R-10225 Yonge St Suite #250, Richmond Hill, ON L4C 3B2, Canada.

Printed in the United States of America.

Acknowledgment

This book is dedicated to my beautiful wife.

Her love and support has made this book possible.

Table Of Contents

Chapter 1

That Which Cannot Be Explained Or Ignored

One of the strongest arguments supporting the existence of unidentified flying objects (UFOs) is the fact that many profound scientific discoveries are initially viewed with skepticism, criticism, and doubt. Subsequently, these advances earn their place alongside the theory of evolution, nuclear energy, and space exploration as legitimate disciplines of science[1]. Throughout the modern history of the 20th and early 21st century, there has existed a large chasm between legacy science and scientific discoveries which adhere to the fringes. The latter has not been considered worthy of academic pursuit until recently, thanks to the advocacy of a short list of individuals. This book will attempt to bridge the gap between legacy science and fringe science and attempt to uncover some common grounds by which both can exist and work together within a realm of acceptable thought and openness to discovery.

Additionally, this is not an attempt to convince the reader of the existence of anomalous activities. It is merely an attempt to shed some light on the multitude of reports and incidents which add to the cumulative evidence surrounding these issues. Professor Avi Loeb, former department chair of

---

[1] KING-HELE, D. UFOs Analyzed. *Nature* 239, 529 (1972). https://doi.org/10.1038/239529a0

astronomy at Harvard, has worked hard to advance this cause over the past several years, emphasizing the fact that it is the inherent responsibility of science and scientists to keep an open mind and avoid the pitfalls of accepting evidence that strictly fits into a narrowly defined narrative.

Even skeptics must admit that genuineness of truth is not dependent upon one's belief system. Hence, we should be open to the possibility of any evidence which comes our way, whether or not it fits our narrative.

Throughout this book, I ask that the reader keep an open mind to the possibility of discovery and think outside of the box. The academic perspective is entrenched within a thought construct which excludes any competing viewpoints that do not fit the narrative of academia, which relies upon a large amount of government funding to conduct research on "acceptable" topics. The question I pose is, who decides what areas of research are acceptable versus which areas are not acceptable? It would appear that acceptable areas of research are tied into funding and encompass motives which strive to secure or support patented and marketable technology.

Among the skeptical group of legacy scientists is the attitude that eyewitness accounts of anomalous events are worthless. However, this author argues that anomalous events are inherently tied to the cognitive experience of the observer. Hence, when particular events are witnessed by tens of hundreds or thousands of individuals simultaneously, science should indeed give credit where credit is due and not dismiss their claims outright. If anything, claims of anomalous events are worthy of scientific investigation and should be taken seriously because, as scientists, we should be open to new ideas, undiscovered physics and physical manifestations which are typically beyond our cognitive abilities to detect.

Perhaps every now and then, a “glitch in the matrix” allows our human senses to see something and/or witness something we typically are not capable of. Indeed, there has been a growing body of legacy scientists who have been able to overcome the cognitive dissonance which is experienced when one's belief system is challenged by new and evolving evidence. This has been observed over the past decade in several subject areas, including the field of UFO/UAP research, lost (or ignored) technologically advanced civilizations, evidence for the existence of non-human intelligence (either terrestrial or extraterrestrial), and alternative theories of the origins of humanity.

The direct research and discussion of these topics, as well as the ancillary issues which surround them, are certainly worthy of serious scientific consideration, which is still in its infancy. To the credit of many early researchers such as Stanton Friedman, Leonard Stringfield, Robert Hastings, Linda Moulton Howe, Bruce Maccabee, John

Brandenburg, and Richard Hoagland, just to name a few, we now have a basis for which academically based scientific inquiry can be directed to both uncover new truths and expand upon the evidence which has already been identified.

There exist an increasing number of enriched hypotheses which support the evolution of microscopic organisms as the precursors of life dispersed throughout our galaxy and perhaps the entire universe (Wickramasinghe, 2014)[2]. Thus, would it be such an insurmountable feat for the evolution of intelligent lifeforms to have occurred within our galaxy and/or the universe? It is now the opinion of an increasing number of both legacy scientists, and those who study anomalous activity that the evolution of such intelligent species is more likely than not.

Furthermore, it is the opinion of many researchers that a long history of visitation and perhaps contact with such beings has occurred throughout time. It seems that a small number of officials in our government are keenly aware of these visits, and every once in a while, they allow some information to leak out regarding this. For instance, in 1968, the class of third-year cadets at the United States Air Force Academy were assigned to read information from their assigned textbook titled, *Introductory Space Science* for their Physics 370

---

[2] Wickramasinghe, N.C., (2014) The Search for Our Cosmic Ancestry, World Scientific.

course. Buried in section 33.6, the conclusion of the author, Major Donald G. Carpenter, proclaimed that Earth has been visited by advanced alien races for the past 50,000 years based on all available data. Furthermore, his conclusion, as conveyed to the students, was that three to four different species of aliens regularly visit our planet.

The textbook was used by the academy from 1968 through 1970 and was abruptly discontinued after the media got ahold of the story and quickly sensationalized the fact that the U.S. Air Force was teaching their students about UFOs and aliens and presenting certain ideas as fact. At the time, this raised eyebrows, and subsequently, the text was discontinued, presumably to allow the Air Force to save face[3].

There is also a growing body of research which supports the possibility of other forms of life within our own solar system. Over the past several years, there have been several published articles in both respectable journals and popular media supporting the possibility of life on Mars, Venus, Enceladus (moon of Saturn), and Europa (moon of Jupiter). Our nearest neighbor, Mars, has commonly been a source of speculation about the potential for life forms and evidence of intelligent life, which may have flourished on that planet thousands of years ago.

---

3 https://thesaucersthattimeforgot.blogspot.com/2022/02/ufos-aliens-us-air-force-academy.html

One of the first scientists to speculate about life on Mars was Giovanni Schiaparelli back in 1877. He based his theory on his astronomical observations of linear formations on the surface of the planet. He assigned them the term *canali,* which, when translated to English, means channels (Launius, 2018: p.248, as cited in magd.ox.ac.uk).

Richard Hoagland and John Brandenburg, PhD, have both presented evidence outlining the case for the existence of ancient structures upon the surface of Mars, similar to that which we see scattered throughout our planet in the form of pyramids and monuments which have the appearance of facial structures. Interestingly, photos of this were first identified by NASA during the 1976 Viking missions upon publication of photos (see figure-1 below), which were taken as the craft orbited Mars.

Over the subsequent years, legacy scientists from NASA have stated that the formation in the photo is simply a trick of light. However, some researchers are skeptical about this explanation considering the other evidence which exists, including pyramidal structures containing angles that are typically not formed by natural processes, particularly on this large of a scale and within a concentrated cluster, such as that which is seen in the Cydonia region of Mars.

Although Richard Hoagland is credited with bringing much of the information regarding Cydonia, including the pyramidal structures and face, he has been the target of much criticism by legacy scientists who cited his paucity of scientific training as grounds for the outright dismissal of his theory and subsequently have labeled him a *pseudoscientist*. Despite these criticisms, Hoagland achieved what he intended to do, which was to raise questions regarding these photographic analyses, which are still being debated today, almost 50 years later.

Interestingly, NASA has never put a rover on or near the Cydonia region of Mars that would settle any lingering doubt, once and for all. One can only speculate why this is not on the list of geographical regions worthy of investigation by the space agency. Another interesting point with regard to skeptic critiques of Hoagland's theory is the fact that a salient point which is often made is the fact that the angles and mathematical anomalies of the D&M pyramid (see figure-2), which Hoagland contends as being significant in his research is often scoffed at by the skeptics as merely occurring by chance or as a result of natural processes.

However, a significant detail which is cleverly ignored by skeptics is the fact that the mathematical measurements would no longer be as precise due to the fact that natural erosion has occurred. Furthermore, we cannot exclude the fact that artificially induced events may have adversely affected the structures in the Cydonia region, such as that which is hypothesized by Dr. Brandenburg in his 2014 paper, which explores the evidence of 129-Xe deposition upon the Martian surface atmosphere as this correlates to similar findings on earth which are only seen after thermonuclear explosions[4]. Based upon these findings, Dr. Brandenburg further speculates on the possibility that a huge thermonuclear explosion occurred at some point in the distant past of Mars in the upper atmosphere near the Cydonia region and another smaller detonation occurred in the region of Galaxias Chaos. Based upon Brandenburg's analysis of the structures in the Cydonia region, he estimates an ancient civilization existed which was fairly primitive, raising the question of the source of these hypothesized thermonuclear explosions. One might consider a rather ominous explanation, such as an attack by a more technologically superior entity. However, this is merely speculation in the absence of supporting evidence.

Much focus over the past few years has been directed toward the idea of uncovering documents, testimony, and evidence which supports advocacy for governmental disclosure regarding the existence of UFOs/UAPs and non-human intelligence. However, one of the goals of this book is to review the various anomalies which all point to the same conclusion and thus render a reliance upon an official statement from governmental officials unnecessary. Hence, it is the opinion of this author that the cumulative amount of evidence is indeed sufficient enough to make that

---

[4] Journal of Cosmology, 2014, Volume 24, Number 13, pages 12229-12280. Release Date, November 20, 2014

judgment call as is. Thus, I would encourage the reader to review what is presented here, which intends to encourage further curiosity such that one would be motivated to research any of the subjects presented further with the goal of rendering your own opinion in much the same way that you would render a judgment as a potential juror. Hence, this requires one to make a determination of both evidence and eyewitness credibility.

Another focus of research is that of lost civilizations (terrestrial civilizations) and the possibility that in the distant past, there may have been a civilization on our planet with advanced technology. One of the most obvious areas of research in this area is the pyramids of Egypt and Egyptology's hypothesis regarding the construction of such pyramids. The evidence presented by academic scientists is that the limestone and granite blocks that comprised the pyramids were dragged, pulled, and pushed into position with mathematical precision down to a fraction of a centimeter and aligned within a north-to-south context to 0.05 degrees of accuracy.

Furthermore, all of this was accomplished without modern construction equipment (supposedly) and to further complicate matters, the average weight of each stone is 2.5 tons with a range of 1.3 tons to 75 tons. Moving blocks of this weight would even be difficult by today's standards, given the fact that we do have capable technology.

However, legacy scientists expect us to accept that this was accomplished by using simple mechanical advantage with counterweights and levers 4500 years ago. Extensive research on this topic has been performed by Graham Hancock, who proposes several deficiencies in mainstream Egyptology theories regarding the construction of the pyramids and proposes alternative possibilities of the construction of the pyramids and the purpose of the pyramids[5] , which academia continues to promote as being constructed as tombs for the deceased pharaohs.

Considering the fact that there is evidence to support similar structures in the Cydonia region of Mars, perhaps a connection exists between these and the Egyptian pyramids. Some researchers have proposed that an ancient civilization which inhabited Mars in the distant past somehow immigrated to Earth as a means to preserve their civilization in light of the fact that the habitability of Mars was quickly becoming impossible. Keep in mind the fact that at one time, 4.5 billion years ago, it consisted of water and a thicker atmosphere. Scientists have proposed that the planet may have flourished with life at that time, including both plant life and organisms. Thus, the possibility of advanced life inhabiting this neighboring world is not so far-fetched.

---

[5] Hancock, Graham (n.d.). "Fingerprints of the Gods - Evidence of Earth's Lost Civilization (1995)."

Furthermore, considering Dr. Brandenburg's hypothesis regarding evidence of thermonuclear blast remnants upon the surface of the planet, a certain narrative starts to emerge. Let's speculate for a moment, shall we?

Let's hypothetically say that our galaxy is the home of intelligent beings, some good and some bad, some who just observe, others who do interact with the native life forms, and some who are opportunistic. Perhaps some of the multitudes of stories we hear about close encounters and abductions by alien beings are indeed true. Perhaps a small number of individuals in government are aware of these facts, which are classified as top secret to the highest degree (for obvious reasons). Furthermore, what if some of this is tolerated in exchange for access to some tangible good or service such as technology, information, or the ability to travel to distant stars and exoplanets within a reasonable time and economy? If any or all of this is true, one easily comes to a conclusion as to why this secret has been kept securely tucked away from the public and remains strictly under the domain of a few individuals commonly referred to in the lore as "secret keepers."

Most concerning is the possibility that ET beings may even be permitted to conduct abductions and, in some circumstances, implant[6] some individuals with tiny monitoring devices so that surveillance of

---

[6] JKLC Productions. (2017). *Patient Seventeen*. United States.

activities and perhaps control over activities can easily be accomplished.

Dr. Roger Leir (1935-2014) was an American Podiatric surgeon who performed surgery on patients to remove anomalous implants within patients. Dr. Leir offered the conclusion, upon analysis, that some of the objects displayed off-world isotopic ratios of elements (such as zinc 64 & zinc 66) and possibly radio frequencies. Although Dr. Leir's research was not replicated, it does offer a possible piece to the larger puzzle of claims regarding alien abductions. I would encourage anyone interested in the subject of alien implants to view the documentary *Patient Seventeen,* which was written and directed by Jeremy Corbell. For these victims, the experience is indeed real, and it often leads to psychological trauma such as PTSD, which requires many to undergo years of therapy. Many, despite this struggle, have to lead "normal" lives.

One of the hallmark abduction cases was that of Betty and Barney Hill, which occurred on September 19th, 1961, while traveling home from Canada in the White Mountains of New Hampshire. This case has all of the elements of abduction, including missing time, a sighting of a UFO, recall of salient events when under hypnosis, and even the reproduction of a "star map" by Betty Hill which depicts the rough outline of the local stars within 50 light-years of our solar system. This reproduction of the local star cluster within 50 light-years of our solar system proved to be quite a remarkable feat by Mrs. Hill, in light of the fact that she sketched it in 1964, which was long before Astronomers had access to the data points which allowed scientific reproduction of this local star cluster. Hence, the fact that Mrs. Hill was able to sketch this in 1964 based on her and her husband's 1961 abduction was indeed strong evidence against this being a hoax event.

Skeptics, including the late Carl Sagan, attempted to explain this reproduction as occurring merely by chance, but the fact remains that several avenues of supportive evidence exist for this case, including circumstantial, scientific, psychological, and sociological. The latter of which concerns the fact that Mr. and Mrs. Hill were a biracial couple who were together at a time when this was very uncommon and was viewed by many at the time as taboo. Thus, the Hills were certainly not inclined to do anything which would warrant any negative attention. As such, one would not expect this couple to participate in a hoax such as this, which could certainly bring unwanted attention from a judgmental public.

A news report appeared on the local Fox News affiliate in southeast Texas Madison County regarding six cows which were randomly killed and mutilated last year. Local witnesses described precision cuts with the removal of the tongue, anus and genitalia along with exsanguination of each animal which apparently occurred prior to the surgical incisions since there was no blood on any of the surrounding surfaces of the animal nor the surrounding ground. Necropsy evaluations by veterinarians involved in these cases consistently have reported dermatological incisions, which are NOT typical of predators or scavengers (Howe, 1997). In fact, it has been documented by ranchers on these various farms where mutilated animals have occurred that the carcasses of these animals go untouched by scavengers such as coyotes, which are typically known to totally strip the carcasses of dead animals. However, in the mysterious mutilation cases, these scavengers do not even come near the dead bodies!

Furthermore, the area surrounding each cow was free of any debris, foot tracks, or tire tracks (figure-2). In the state of Texas, as in most states, this is a felony offense. However, in the vast majority of cases, the perpetrators are never found. These "cattle mutilation" cases have been occurring since the 1970s and have been within the wheelhouse of some UFO investigators due to the fact that many witnesses in the geographical areas where these incidents occur have reported seeing strange lights in the skies in and around the time of when these incidents have been estimated to occur. Agricultural experts and some veterinarians who are familiar with these cases have stated off the record that feats of this nature would take a significant amount of planning, personnel, and expertise to accomplish without being detected. Furthermore, accomplishing this task in the dead of night and for no clear motive simply does not make any sense.

These are widespread occurrences which have spanned the country, including the states of Colorado, Texas, Oregon, Pennsylvania, Nebraska, Montana, Idaho, California, Kansas, Wisconsin, Minnesota, and Illinois. The New York Times reported on these incidents on October 30th, 1975, noting a large cluster of incidents in 10 states all sharing the same modus operandi noting bloodless surgical excision of cattle eyes, ears, udders, anus/sex organs, and tongues of numerous cattle. It was initially blamed on "cult" activity; however, the paucity of evidence in all of these cases has never led to any prosecutions, which would explain the vast majority of these cases which appear to occur at random. A federal investigation by the Federal government ATF bureau failed to find any evidence linking cultist groups to cattle mutilations, and as a result, the case was closed. Almost fifty years later, these cattle mutilation incidents continue to occur still without any apparent explanation.

Just this past week, the Congress of the United States held the second set of congressional hearings on UAP/UFOs, whereby three witnesses (under oath of perjury) were questioned by various members of Congress regarding the existence of extraplanetary vehicles, crash retrievals[7], and recovered alien bodies. The most shocking testimony was provided by Luis Elizondo, who point-blank answered a responding affirmative when questioned by Congress about the

---

[7] The recovery of crashed or downed extraterrestrial vehicles by the U.S. military.

existence of the aforementioned topics. Most salient about his background is the fact that he is a former military intelligence official who spent 10 years managing a Pentagon-sanctioned UAP program. This hearing comes on the heels of the first hearing, which occurred in July of 2023, where Retired Major David Grusch became the first government official to testify under oath that the United States has recovered alien vehicles, alien biological materials (bodies), and has a robust reverse engineering program which is subcontracted to various defense contractors.

Never in the history of UFO research has something as profound as congressional hearings in front of sitting members of Congress taken place. However, the citizens' hearing on disclosure can be given credit for putting the most recent congressional earrings in motion. This hearing occurred in 2013 and involved testimony from leading researchers, ex-government officials, and witnesses given to retired members of Congress and the Senate.

We will cover some of the historical highlights of the UFO/UAP phenomenon in subsequent chapters, which really, in terms of modern history, achieved widespread media reporting with the Roswell incident which occurred in 1947 (see newspaper article figure-4). This article, which was published immediately after the incident, gained widespread coverage in major newspapers across the United States. The next day, the Air Force claimed that the original report was false and that this downed object was nothing more than a “weather balloon.” Subsequent research into the incident by, most notably, Stanton Friedman and Don Berliner,[8] uncovered ample official and unofficial evidence that the original article was indeed correct and implicated the powers that be in an exquisite cover-up, which continues to this day, that indeed what crashed outside of Roswell, New Mexico was an off-world vehicle piloted by non-human intelligent beings.

The craft was recovered by military personnel and transferred to top-secret military facilities in the southwest U.S. and to Wright-Patterson Air Force base near Dayton, Ohio. This incident paved the way for the formation of a group of 12 individuals who were top military brass scientists, tasked with both leading a new program designed to collect and examine the artifacts recovered from this and later incidents involving downed alien craft.

Furthermore, a program was eventually established with the goal of reverse engineering the technology. This group, known as MJ-12, was the highest of top secret and was formed by and responsible for reporting directly to President Harry Truman. Some researchers believe that as the years went on, the supervision of MJ-12 was eventually subtracted from the president’s purview and committed entirely to that of the military-industrial

[8] Recommended reading: “Crash at Corona” published in 1992, 1997 by Stanton Friedman & Don Berliner

complex, where it remains to this day. This system of compartmentalization with access to only a few individuals with top-secret clearances who are "in the know" is how this secret is maintained.

There is speculation that maintenance of this information probably involves financial security for those tasked with guarding this information at the highest levels. But don't take my word for it. I encourage the reader to keep an open mind while reading this book and to do their own research using some of the sources I've highlighted in the figures and footnotes and decide for themselves.

Figure-1. NASA Viking photo of the face on Mars.

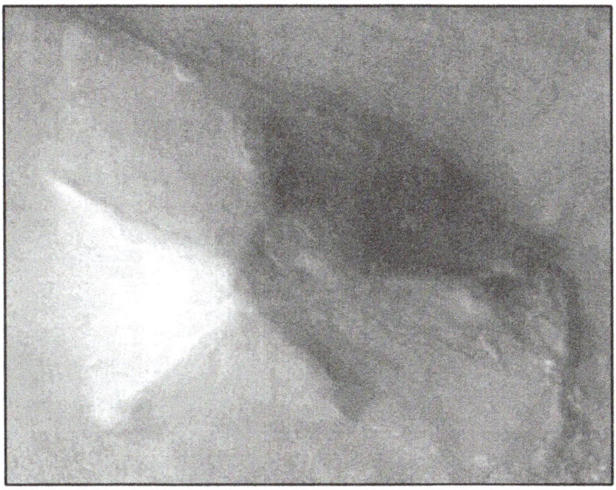

Figure-2. NASA photo, D&M Pyramid.

Figure-3. Cattle mutilation incident reported in the Arizona Daily Sun, March 6th, 2020.

Roswell Daily Record

RECORD PHONES
Business Office 2288
News Department
2287

Roswell, New Mexico, Tuesday, July 8, 1947

5¢ PER COPY

RAAF Captures Flying Saucer On Ranch in Roswell Region

Claims Army Is Stacking Courts Martial

Indiana Senator Lays Protest Before Patterson

House Passes Tax Slash by Large Margin

Defeat Amendment By Demos to Remove Many from Rolls

Security Council Paves Way to Talks On Arms Reductions

No Details of Flying Disk Are Revealed

Roswell Hardware Man and Wife Report Disk Seen

Ex-King Carol Weds Mme. Lupescu

Figure-4. Original newspaper article reporting on the recovery of a flying saucer in Roswell, New Mexico July 8th, 1947. Source: https://en.wikipedia.org/wiki/Roswell_Daily_Record#/media/File:Roswell_Daily_Record._July_8,_1947._RAAF_Captures_Flying_Saucer_On_Ranch_in_Roswell_Region._Full_front_page.jpg

Chapter 2
Notable UFO Sightings

Throughout this book, I have tried to incorporate reports of various UFO/UAP sightings into the context of the verse to give color and provide context to the meaning of the passage. In this chapter, we will concentrate on specific cases which provide context and help give a defined structure to the subject of ufology.

Certain aspects of the cases presented within this chapter may resurface in subsequent chapters. The purpose of this is twofold. First of all, there is always some additional aspect of a specific case which makes its way into a particular UFO/UAP-related context or in peripheral relation to the UFO/UAP subject. An example of this is when we discuss particular guidance and/or regulations related to the military or the FAA. Secondly, my view of the subject of ufology requires an academic discussion because of its detailed history and requirement of knowledge within the STEM (Science, Technology, Engineering, Mathematics) disciplines. Thus, the purpose of providing this information to our readers is to both satisfy the reader's interest in the topic and also to educate them about the topic which we view as an evolving area of scientific investigation.

It is important for the reader to understand that the topic of UFOS/UAPs is often broad, obscure, and sometimes encroaches upon other subject areas, some of which include discussions regarding non-human intelligence, abductions, cattle mutilations, implantations, crop circle formations, religion, and alternative theories of human origin. Thus, repeating aspects of particular UFO/UAP cases is often unavoidable to provide the appropriate context for said topic.

Moreover, there exist additional implications from historical, political, religious, economic, and military perspectives, which require emphasis on certain cases, and thus repetition becomes occasionally necessary. Renowned researchers Coral and Jim Lorenzen, founders of the Aerial Phenomena Research Organization (APRO), which was in existence from 1952-1988, put forth the following statement regarding the existence of only two choices of theory about UFOs:

1) The objects and their operators are physically real. The remaining speculation concerns their identity and origin and, eventually, motivation.

2) The population of this world is falling victim to a particularly insidious and apparently contagious mental disease which generates hallucinations involving specific types of airships and humanoids.

Before we move forward, it is important to understand first and foremost that because of the mountain of evidence which has accrued over the past century, one can be confident that UFOs are indeed real, physical objects with biological occupants that have been actively visiting our planet for (most-likely) hundreds, if not thousands of years. Much of this visitation appears to be consistent with active surveillance of our activities as a species.

Trent Case

Any subject area which relies on scientific analysis often relies on certain truths or precedents against which all subsequent findings are measured. Within the field of Ufology, many researchers regard the Trent flying saucer sighting and photos to be the epitome of this precedent. This sighting occurred on May 11th, 1950, on a farm located near Sheridan, Oregon. A married couple, Paul and Evelyn Trent, were going about their daily routine, which involved feeding the rabbits on their farm, when suddenly, the wife Evelyn was startled by a silent, disc-shaped craft hovering over their farm. Luckily, her husband Paul was able to quickly snap two photos, which are, to this day, considered to be possibly the best photos of a flying saucer (see figure-1).

At the time of the sighting, the Trents had to deal with much criticism and skeptical analysis, and also harassment from unknown government agencies, who even insisted upon searching their home and property. Throughout all of the 45-plus years which transpired after the photo was taken, the Trents never embellished or changed their story; it was always consistent.

Renowned physicist Dr. Bruce Maccabee investigated the Trent photos and employed various methods to examine the photos, such as enlargement of the photos to inspect for any attachments to the object which may have been used to control or suspend the object in the air. However, none were found. Furthermore, some shadows in the photos were called into question by skeptics, trying to cast doubt about the time of day the photos were taken. The Trents reported that the photos were taken at sunset, but the skeptics were trying to force the idea that the photos were taken in the morning. Dr. Maccabee's extensive analysis of the shadows ruled out any optical fabrication and concluded that the photos were not taken in the morning.

Furthermore, Dr. Maccabee concluded that the object was at least one kilometer away from the Trent's property at the time the photos were taken. That distance would be difficult for anyone to hoax, especially without being noticed. Dr. Maccabee concluded the photos were authentic and corroborated fully with the claims reported by Mr and Mrs Trent.

In an interview that occurred in the 1990s, Evelyn Trent describes the flight characteristics of the saucer as being void of any flight surfaces and also confirms the fact that upon departure from its hovering mode, the craft tipped on its side and sped away faster than anything she had ever witnessed in her life. The fact that she described this tipping or rotating maneuver is consistent with reports which have been received from military pilots who have witnessed anomalous craft while on training exercises off the coast of Southern California in 2004 and is also consistent with Bob Lazar's description of the operational modes of the craft which he worked with. Hence, when these craft are in a hovering mode, the antigravity emitters which are responsible for the craft's propulsion, are in a low power mode, which he terms the Omicron configuration.

Conversely, when the craft's emitters are configured into the Delta configuration, a high power mode is achieved, which is used for quick travel either within the atmosphere or interstellar travel. Probably the most notable aspect of the Trent case is the fact that these people of modest means never sought to achieve fortune or fame because of these photos. In fact, they were reluctant to come forward and regretted the fact that they had done so later on in life.

During the interview, they made it a point to state that if they could have a do-over, they would never have taken the photos in the first place because they have caused nothing but trouble for them.

Figure-1. Trent photo. Source:
https://en.wikipedia.org/wiki/McMinnville_UFO_photographs#/media/File:McMinnville_UFO_1.png

It is also important to realize the *"ours versus theirs"* argument that has been put forth by various researchers, most notably the renowned researcher Michael Schratt[9] , who has been investigating the topic of UFOs for the past 25 years. This concerns the fact that a great deal of sightings are actually reverse-engineered craft. Lending support for this argument is the fact that eyewitness accounts often include details of the craft which include the presence of tubes,

---

[9] https://www.youtube.com/watch?v=bKswZJfJdAA

cylinders, hardware, and rivets on the bottom and/or sides of the craft when seen from underneath.

Conversely, historical accounts of what is considered true extraterrestrial craft are typically void of these details and are described as smooth and seamless. Oftentimes, the skin of the true ET craft is described as almost the appearance that they have been dipped in wax or, more recently, subject to 3D printing such that all of the body/skin surfaces are continuous throughout the entire craft.

These same features have also been described by scientist Robert Lazar, who has reported his participation on a reverse engineering team tasked with working on the propulsion system of a recovered disc in the late 1980s at a top secret military facility in Nevada. The extra details added by Lazar, however, are that these smooth and seamless topical features are not only present on the outside of the craft, but they are also the prime feature of the surfaces on the inside of the craft, including control panel surfaces and occupant seats. Lazar's claims have been vetted by renowned UAP researchers and the consensus at this juncture is that his claims are credible. This has become even more apparent most recently with the U.S. congressional hearings, where allegations of the lengthy existence of reverse engineering programs have been brought to the attention of the U.S. inspector general and members of Congress under oath by a whistle-blower, David Grusch.

One common theme which is continuous within the vast history of UFO/UAP cases is the fact that whenever a credible sighting, crash, or contact with vessel occupants occurs, it is subsequently met with an intense clean-up and debriefing by military retrieval teams, in conjunction with local law enforcement and emergency services. However, the latter personnel are often kept in the dark as much as possible and are only provided as much information as needed to assist in clean-up efforts. For example, in August of 1952, a case which occurred in Ely, Nevada,[10] involved a downed vessel that was ultimately discovered by some of the citizens of the small town, and the legend has it that all of them suspected this was an ET craft because of the fact that sixteen deceased humanoid bodies were found in and around the craft.

Military crash retrieval teams were quick to cordon off the area and the typical warning of residents not to speak of the incident was delivered by these officials. UFO/UAP literature documents quite similar scenarios in the majority of crash investigations and credible sightings whereby crash retrieval teams quickly seal the area, clean up any debris, and "counsel" any witnesses. One must understand that when confronted with such a spectacular situation, most people inherently will follow the directions of officials out of fear. This was

[10] The Ely, Nevada UFO Crash of 1952. Published April 30th, 2011. ufodigest.com.

even more true back in the 20th century before the evolution of the internet and the mass communication network which has evolved with social media.

Nowadays, people are much more apt to upload this information in real-time, often before officials have even had the chance to respond. A recent example of the ability of a citizen to immediately upload a credible sighting onto the internet happened over the skies of New York City on March 25th, 2024, when a commercial airline passenger was looking out the window and filming with her smartphone. She was able to capture the image of a flying cylinder. This was quickly reported to the FAA, and a civilian UFO organization has also reviewed the footage, which appears to be authentic.

The object has all of the characteristics associated with UAP authenticity, including the absence of typical aerodynamic craft, such as wings, tail-section, rotors, exhaust plumes, etc. The speed, altitude, and proximity of the object to LaGuardia airport seem to point away from this as being a simple drone, according to civilian investigators.

Along with social media, major news outlets also carried the story, including the New York Post, NewsNation, and the Daily Mail. As is true with most sightings, there's been no further follow-up, and no obvious statement could be found on the FAA website regarding the incident, despite the fact that it was reported that the incident was reported to the FAA when it first occurred. This is concerning since this occurred above LaGuardia, and thus, one would think there's an obvious risk to air safety, as well as the fact that in the images obtained, the object certainly seems to be fairly close to the aircraft from which the photo was taken.

The O'Hare Incident

The paucity of responsiveness by the FAA with regard to incidents of airspace incursions, unfortunately, is a common occurrence, even in the most serious cases where the safety of flight could be an issue. One of the most notable cases involving flight safety risk occurred 18 years ago on November 7th, 2006, at 4:14 pm at one of the busiest airports in the world, Chicago O'Hare. This incident occurred over gate C17 and was witnessed by 12 employees, including gate workers, pilots, flight attendants, and baggage handlers.

The incident was reported promptly to both the tower officials and the FAA. During this time, Jon Hilkevitch was the transportation reporter for the Chicago Tribune and he reported that this "metallic appearing disc" was reported to have hovered above gate C17 for around five minutes. Furthermore, it was also reported that there was much communication over the airwaves between tower personnel and other airport employees at the time. Interestingly, the tower personnel denied seeing the object and one official can be heard over the radio saying that even if he did see anything of that sort, he probably would not have reported it regardless.

Luckily, there were willing employees who did file a report with the FAA however, the resulting investigation certainly did not vindicate what the witnesses observed. The FAA basically refused to pour any major resources into investigating the incident in spite of the fact that there were a dozen witnesses all reporting the same physical craft.

In the end, officials simply port forth the concluding theory that the O'Hare incident was due to a "weather phenomenon." Astronomer Mark Hammergren suggested the idea that this was a "hole punch cloud," and this was enough to taint the integrity of this event despite the fact that this type of weather phenomenon occurs when the temperature is between 14 to 36 degrees Fahrenheit. The temperature at O'Hare that particular day at that particular time was 56 degrees Fahrenheit. In fact, the entire day of November 7th, 2006, the temperature range was from 43 degrees to 58 degrees Fahrenheit, which is obviously much too high of a temperature for a hole punch cloud to successfully form.

The witnesses and anyone who was informed about this incident realized that this was still the immediate post-9/11 era, where any possible threat to flight safety should have been taken seriously by officials. However, the magnitude of response that was happening in this case was concerningly dismal. FAA officials simply discounted all of the reports and concerns of the airport employees. The field of UFO/UAP research views this particular case as a prime example of how this topic is quite often not allotted proper attention for reasons which are largely unknown.

Triangular/V-Shaped Craft

Close to 3,000 sightings by extremely credible witnesses in the 1980s and 1990s took place in the Hudson Valley, NY region. Largely described as a boomerang and V-shaped craft, more than 5,000 individuals witnessed this string of sightings. Witnesses included law enforcement, engineers, corporate executives, doctors, and other highly credible professional people. The principal investigators of the Hudson Valley cases were J. Allen Hynek, Philip J. Imbrogno, and Bob Pratt. Their research was reported in their book, *Night Siege, The Hudson Valley UFO Sightings*. These authors reported these sightings occurred within a 1,400 square-foot mile region which encompassed the following New York State counties: Dutchess, Putnam, and Westchester and also the following Connecticut counties: Fairfield, Litchfield, and New Haven.

A small portion of these sightings were the result of conventional aircraft flying in V-shaped formation. However, the vast majority could not be identified and even included cases where the craft approached the witness within 500 feet. In some of these close encounter cases, the object acted aggressively in terms of its approach and blinding spotlights which it directed towards the witnesses. Some of the witnesses felt that the object's presence was so intense that they felt as if it was trying to communicate with them. Witnesses described the spotlights as being so intense that they had to close their eyes.

Interference with car ignitions and radios has also been consistently reported in many close contact cases. This seems to be a common theme in cases, particularly where witnesses are in vehicles at the time of the close encounter. This electronic interference is largely attributed to the electromagnetic properties of these objects.

Other sightings involving V-shaped craft have occurred in Belgium, Southern Illinois, and Phoenix, Arizona, throughout the 80s, 90s and early 2000s and were witnessed by large numbers of individuals, especially the Phoenix lights incident which was witnessed by thousands. The common thread which links these cases, as well as the other genuine cases is the fact that the craft traversed through the atmosphere silently and had no visible flight surfaces or means of propulsion. Perhaps the silence of this craft can be explained by the electromagnetic properties, which one can assume might be a byproduct of the antigravity propulsion mechanism described by Bob Lazar.

If one is to assume this is accurate, then during normal operation, these craft are, in a sense, operating within an artificial gravity bubble rather than directly in contact with the atmosphere. This may explain the ability of the craft to operate at tremendous speeds which would typically annihilate any onboard passengers. However, the protective bubble might provide an effect by which inertia is inconsequential upon the physical body, thus negating the possibility of injury or death.

As one can see, there seems to be an assortment of crafts which can be V-shaped, saucer shaped, cylinder shaped, circular shaped and even two or more shapes within each other, such as a cube within a sphere. There are also differences with regard to reports of the interior or crafts which have been inspected in crash retrieval cases. For instance, the vehicle described by Bob Lazar has an interior that is void of any buttons or control panels. However, the vehicle described in the Aztec case (at least that which was originally reported by Frank Scully in his 1950 book *Behind The Flying Saucers)* did, in fact, have interior features consistent with buttons and control panels. As a matter of fact, Scully makes it a point to note that the eyewitnesses thought about trying to press some of the control panel buttons just to see what would happen. But after debating about this, they ultimately decided against this because of the risk of activating something on board the craft and then not being able to turn it off.

Furthermore, one would think that the eyewitness would be apprehensive about even entering the craft, as there would be a perceived risk of getting locked inside. Perhaps this is one of the reasons why first-hand witnesses were so quick to turn over the scene to the military officials who arrived because of the unknown origin of the craft they had come into contact with.

One could assume the apprehensiveness which some witnesses probably felt upon stumbling onto these craft. Common sense would allow one to make the assumption that when witnesses are first to arrive on the scene, much caution is exercised out of concern for incident or injury. Furthermore, it is safe to assume that most people are just trying to get on with their daily lives, and when something extraordinary is witnessed, most level-headed individuals will devote no more time to the incident than is absolutely required to enable them to resume their normal daily activities. Hence, the appearance of either government officials or military officials is often met with a sign of relief for witnesses who just want to get back to their jobs, families, and responsibilities. Most people want little disruption to their routines than is absolutely necessary. Thus, when an official advises them, "there's nothing to see here," most of these witnesses will comply and go about their business.

Renowned researcher Stanton Friedman spoke extensively about that which he framed, "category four events" as being valid from a scientific standpoint. These events are unpredictable with regard to *when* they will occur but occur with regularity and oftentimes go unobserved directly. This includes events such as plane crashes, car accidents, and crimes. Most often, for these events we must rely upon eyewitness testimony, assessment of physical injury, and/or property damage.

UAP/UFO sightings most often fit into this category. The reliance on eyewitness accounts of UAP/UFO events is particularly troubling for certain individuals, most notably pilots. Traditionally, pilots who provide official reports of UFO sightings have been met with intense scrutiny and even risk of career interruption or loss. This practice originates from *Project Grudge,* which began the practice of relieving pilots of duty who reported a UFO encounter and subsequently forced them to obtain a psychological examination to ensure they were safe to fly aircraft. Turns out, this was a surefire way of preventing pilots from filing official reports because pilots were indeed fearful that they could lose their careers, pensions, benefits, etc.

This same scrutiny and stigma persists today among the aviation community. However, there are efforts towards changing this practice, especially since it has been brought to light during the recent congressional hearings on Capitol Hill in 2023 and 2024. Still, most pilots will only provide unofficial accounts of UFO encounters, and when they do feel the need to report such instances, they often file their reports with civilian agencies, such as MUFON[11]and/or NUFORC[12], anonymously.

---

[11] MUFON: Mutual UFO Network.

[12] NUFORC: National UFO Reporting Center.

Legacy scientists within the world of academia have traditionally relied upon laboratory experiments where confounding variables can be controlled, and the direct effect of the independent variable upon the dependent variable is analyzed quantitatively. Despite this controlled environment, the proof is only demonstrable through multiple replications involving the same or similar experimental techniques, and oftentimes, there is still a lingering question of causality due to the fact that the dependent variable is most often *not* fully affected by the independent variable. There are always extraneous variables at play. For instance, in the pharmaceutical industry, when new drugs are tested upon subjects, multiple variables exist including variations in body chemistry, metabolism, genetics, absorption, chemical or drug interactions, and placebo effects. Thus, researchers can hardly say with 100% certainty that a drug is fully responsible for its effects on an organism.

Years of research are needed to gather enough evidence to show benefit. A most recent example of this is that which occurred with the COVID and influenza vaccinations. Despite the fact that these vaccines are often reformulated to keep up with changes in viral mutation, there is never complete effectiveness and as a result, a significant number of patients still will contract the disease.

Chapter 3
UFOs Within The Context Of Military, Economics, And Government

UFO/UAP research has demonstrated a pattern by which reports of sightings, contacts, and crashes of anomalous objects increased exponentially after the United States began its nuclear testing program back on July 16th, 1945, with the Trinity test. Make no mistake, prior to this, reports have been documented throughout history. However, reports were much more infrequent compared to after the advent and testing of nuclear weapons. Perhaps the testing of nuclear weapons created a defect in the space time continuum which created a ripple effect that was detected by our cosmic neighbors. Or perhaps if the fourth-dimensional hypothesis is correct, UAPs travel into our three-dimensional spaces through preconstructed conduits, which also conduct the immense energy produced from the splitting of the atom during the detonation of a nuclear device. Perhaps this is interpreted as a threat to the fabric of the universe, and as such, they responded in droves to our small rocky world to find out exactly what we were up to and perhaps in an attempt to put the brakes on whatever havoc we were in the process of creating with this unchecked manipulation of atomic power.

This history of manipulation of the atom culminated with the final test on September 23rd,

1992, in the top-secret deserts of Nevada, a long-awaited finale after the 1,032 previous nuclear tests which took place in that 47-year span. Perhaps it is not surprising that the most compelling crashed UAP cases, contact cases, and credible sightings occurred within that period. Moreover, in light of our less-than-stellar track record of war, death, and destruction as a society, is it any surprise that our cosmic comrades would rather observe us from a distance and interact with us infrequently and only when they can procure and maintain an upper hand advantage of control over us, as they can truly never be too sure what we are capable of.

One must keep in mind the context here, which includes the fact that the United States holds the record as the only country in history that has ever used a nuclear device against another country, resulting in the deaths of 105,000 people and the injury of an additional 94,000 people. Perhaps this ominous record has earned us a permanent place in the cosmic probation program, which requires constant surveillance, regulation, and monitoring from our cosmic probation officers.

The bulk of the research with regard to UAPs which have disabled the launch capabilities of surface-to-air missiles in the United States and former Soviet Union, is credited to Robert Hastings. During the Cold War, it was quite difficult to procure this information from sources behind the Iron Curtain. However, with the dissolution of the Soviet Union on

December 26th, 1991, it slowly became more progressive in terms of the exchange of information regarding UAPs, and as a result, it was discovered that many of the same phenomena that were being experienced in America were also being experienced in the Soviet bloc. This includes credible sightings, hardware recovery, and interference with nuclear missile launching facilities. It is important to note that these similarities were discovered not through official channels but through unofficial channels, for the most part. However, procurement and exchange of the information was accomplished through credible individuals such as former government officials and military personnel.

In a classic series of Larry King Live episodes back in 2008, Mr. Hastings appeared along with Robert Salas, who was a former officer at Malmstrom Air Force Base in Great Falls, Montana and bore witness to missile disabling capabilities of UAPs in the 1960s. Also joining this interview was Robert Jacobs, PhD, who formerly performed technical photography at the Vandenburg military base from 1963 to 1966. During one missile launch, Mr. Jacobs unknowingly recorded a video of a UAP seamlessly neutralizing a surface-to-air missile with (what was later determined) to be some type of plasma or laser beam. He was confronted with this revelation after the fact when presented with this video evidence by his supervisor and members of an undisclosed three-letter government agency.

After debriefing Mr. Jacobs on the incident and concluding that he had no nefarious involvement in the incident, he was given instructions to never speak of the incident again. Interestingly, during this Larry King episode, Bill Nye was brought into the discussion as the skeptical wing of the debate and provided alternative conclusions which appeared to irritate the witnesses, and the ensuing discussion thus became quite heated at times. Although some may view this as unprofessional, one must consider the fact that debates between skeptics and researchers have been occurring for the past 80+years with regard to the UFO/UAP subject and signifies the fact that even though it leads one to wonder about the existence of extraterrestrial life (and whether or not this life is actually visiting our planet), the subject itself is invariably human insomuch as it stirs up emotions of fear, anger, and frustration.

Fear in the sense that humans are inherently fearful of the unknown, and anger/frustration in the sense that the thousands of witnesses to anomalous events are often criticized and ridiculed by skeptical individuals who are constantly trying to "make sense" out of our world, which is clearly not completely understood.

With the proliferation of nuclear weapons came the simultaneous proliferation of top-secret military installations, particularly in the southwest United States, including areas in Nevada, Arizona, and New Mexico. Thus, it should come as no surprise that a great deal of credible UFO/UAP sightings and contacts originate in the vicinity of these bases. For example, a well-documented case reported to the National UFO Reporting Center[13] involves a fast-moving disc that flew over the top of four military crafts (A-10, F-16, F-15E & F-22) which occurred during the national anthem of a race at the Las Vegas motor speedway on Sunday, March 3rd, 2024. This was recorded on video which has been posted on the NUFORC website. Not surprisingly, the close proximity of Nellis Air Force Base, which abuts the top-secret Nevada test sight, "aka" Area 51, is well within range.

Another notable case reported to NUFORC with video evidence involved a blue orb caught on video over the base of Sunrise Mountain in Las Vegas, Nevada, which appeared to display characteristics of genuine UAP, including sharp, erratic movements, the ability to quickly avoid military aircraft which at the time tried to pursue this object, rotational movements, and interaction of a plasma-like phenomenon within the earth's predominantly nitrogen atmosphere. Upon close inspection of the video, it seems possible that this apparent sphere is enclosing an inner cube-like structure.

This type of UAP design has also been documented by the military pilot Lieutenant Ryan Graves in 2014 when his military (Navy) unit flew

---

[13] NUFORC.ORG

training exercises in the Virginia Beach area. Mr. Graves has even testified under oath to Congress regarding his encounters with these UAPs.

It is not surprising that crash-retrieval "teams" are able to respond to downed objects very quickly and efficiently, considering the proximity of many top-secret facilities in the deserts of the American southwest. Furthermore, these facilities typically house hundreds (if not thousands) of military personnel. In fact, it seems that any place which is infused with military-grade weapons has a tendency to also attract our cosmic overlords, which again goes back to the possibility that we are being monitored and regulated to a certain extent. As we have clearly not progressed beyond the primitive instinct of conflict and warlike tendencies to achieve the demeanor of a civilization beyond type zero, if we extrapolate the Kardashev scale to encompass not only a civilization's energy-harnessing ability but also the ability of its members to coexist peacefully with each other.

This would require our abandonment of both individual and societal tendencies towards conflict and war. This, unfortunately, is a goal that often seems beyond our grasp, as a significant part of the economy thrives upon military preparedness for battle. Hence, with respect to our great President Eisenhower, the military-industrial complex continues to thrive and provides the basis for a significant chunk of our economic resources. In his

January 17th, 1961 farewell address, President Eisenhower made the following statement:

> ***In the councils of government, we must guard against the acquisition of unwarranted influence, whether sought or unsought, by the military-industrial complex. The potential for the disastrous rise of misplaced power exists and will persist. We must never let the weight of this combination endanger our liberties or democratic processes. We should take nothing for granted. Only an alert and knowledgeable citizenry can compel the proper meshing of the huge industrial and military machinery of defense with our peaceful methods and goals so that security and liberty may prosper together.***

-President Dwight D. Isenhower

To provide a bit of perspective as to the context of the above quote, the amount of reported funding to United States defenses for fiscal years 2019 through 2022 was 734 billion, 778 billion, 806 billion, and 876 billion, respectively. Keep in mind that this is just the reported funding. Each year, the Congressional Budget Office seems to uncover discrepancies in the budget such that vast sums of monies go mysteriously missing. In fact, The Pentagon, aka the Department of Defense, has failed routine audits six times and never actually completed an audit until 2018. These failed audits have estimated 1.9 trillion dollars in unaccounted money during the current fiscal year 2024 alone. This type of mismanagement would never be tolerated in any other government agency. There would be congressional hearings, firings of those in prime leadership roles, and possible indictments. But sadly, nothing is ever done about this problem and to make matters even worse, congress routinely votes to approve increasing amounts of money each year. This begs the question, where is all of this money going?

To sum this up, we are spending hundreds of billions of dollars allocated to the DOD each year, and a sizable portion of that money goes unaccounted for. Yet, we are seeing stories in the news media outlining the fact that we cannot fully protect our own air bases. For example, an article appeared in eurasiantimes.com titled, ***"F-22 Raptors: U.S. Seeks Anti-Drone Nets to Protect Its Stealth Fighters; Mysterious UAVs Swarm Langley AFB"*** on October 14th, 2024. The most concerning part of this article mentioned was the fact that military officials were unable to determine the source of these incursions! One would expect that the amount of money the U.S. taxpayer is spending each year on national defense would virtually guarantee complete protection of our skies and military/air bases (on our own soil), but this is apparently not true. This raises a multitude of concerns but unfortunately, radio silence has been the status quo from our government officials.

One might speculate that a significant portion of this funding is routed into top-secret programs. Some UAP researchers have even proposed that some of this money is being spent on reverse engineering programs of alien technology. Still, other researchers have presented evidence of a top-secret space program, the details of which are hidden from the American public. In fact, some researchers[14] speculate that American astronauts have already set foot on the surface of other rocky planets and moons within our solar system, such as Mars, Titan, Europa, and Ganymede.

Moreover, some speculate our travels could have even breached the bonds of our solar system and traveled to some of our closest neighboring stars to explore exoplanets such as Proxima Centauri, Tau Ceti, Wolf 359, Barnard's star, and Gliese 12B. If this is true, this would mean the harnessing of alternative propulsion methods and/or traversing of accessible wormholes has already been achieved by mankind. This monumental achievement could certainly explain the unchecked defense spending and routine discovery of budget discrepancies which are routinely brushed under the proverbial rug. If any of these suspicions are true, it is this author's opinion that withholding this information from the public, STEM professionals and academia certainly skates along the

---

[14] Richard Dolan has lectured and written about the possibility of a secret space program extensively

margin of ethics in a society that touts itself as technologically based.

We have a situation where particular technologies, including fiber optics, quantum technology and touch screens, have no clear-cut evidence from origin to completion during the last 100 years. This also begs the question of how the rapid technological advancement evolved in just over 200 years, when previously to that, 12,000 years of advancement was slow, often with thousands of years passing before major inventions were made (e.g., use of controlled fire, bow and arrow, hand-held tools, etc.). I am not suggesting any conclusion to these questions. I am just pointing out some discrepancies in the data when viewed from a historical perspective.

It is the opinion of this author that one cannot separate completely the mission and intent of government which takes the form of public administration from the subject of UAP. 65 years ago, the Brookings Institute[15] suggested the government take pause before deciding upon the extent to which official acts of disclosure should actually go forward. Calling into question the public reaction from a position of religion, culture, and society could potentially have devastating repercussions for the maintenance of societal structure. Hence, if provided with information that higher powers than the

---

[15] https://www.brookings.edu/wp-content/uploads/2014/05/space_extraterrestrials.pdf

government officially exist (i.e., technologically advanced extraterrestrial beings), there exists the possibility that significant upset in society structure can result, including but not limited to reluctance to engage in economically productive activities. Hence, withholding upsetting information is a method by which the government avoids the "tragedy of the commons," which is a situation in which some individuals act in a selfish manner and take advantage of public resources.

We saw this recently during the pandemic when looters and rioters attacked neighborhood shops and took over entire parcels of land, setting up camps and disregarding the authority of police officers. In order to understand this fully, one must realize a basic tenet of government is that it exists in a constant state of flux regarding decisions and often, decisions are made with vague generalities which are left open to broad interpretation by administrators and those who hold political power. Unfortunately, decisions and the delay of decisions are decided not only on the basis of the public good, but are also decided upon the basis of survival of the individual politician in the next election cycle. Hence, those representatives whose constituents comprise a significant number within their district working for military contractors, for example, may be influenced to vote a certain way on a bill that mandates governmental disclosure of certain facts which are classified. Furthermore, the compartmentalization of government has created a situation where plausible deniability is the status quo, especially when it involves programs that require billions of dollars from taxpayer-funded coffers.

Hence, public officials may be acting in the best interest of the public toward the end of maintaining an ordered societal structure, which in and of itself works toward a goal of national technological advancement. This is far from the writings of Walter Lippmann, whereby public interest in its purest form is presumed to be based upon the concepts of benevolence and without the intent of personal gain[16].

Conversely, one can also argue that the secret keepers are indeed engaging in acts of personal gain by holding information from the public for which it has the right to know. However, as we can see, the structure of society, government, and institutions holds that power is concentrated on a few individuals who collectively decide what the masses can safely be given access to without upsetting the balance of societal order. Perhaps we have not evolved to the point by which complete governmental disclosure is safe in terms of maintaining this order. Unfortunately, we see examples of this disorder every day with crimes against individuals, institutions, private enterprises, and crimes against humanity (e.g., The Holocaust). Hence, one must take into account the maintenance of public order versus the risk of disorder when deciding what information is

---

[16] Walter Lippman (1889-2004) Journalist and political philosopher who critiqued foreign policy and presented analysis of public opinion in his writings.

appropriate to disclose to the public. I venture to say this is true not only with topics such as UAP/UFO disclosure but also with an assortment of other topics which can result in significant societal structure interruption. This would include anything from mass casualty events to natural disasters and beyond.

A well-documented example of an event which warranted far more disclosure than has been offered to the public since its occurrence almost 28 years ago was the March 13th, 1997, Phoenix Lights incident which was witnessed by thousands of Arizona residents. This incident garnered international attention and a paucity of explanations have been provided by government officials, for the most part just blaming this mass sighting on military flares. However, detailed investigation of this mass sighting has consistently revealed corroborating eyewitness accounts from credible witnesses, including police officers, pilots, and physicians who all witnessed the same or similar objects which displayed identical flight characteristics in terms of flight dynamics (speed & altitude), vehicle shape (V-pattern), duration of flight[17], and silent motion of the object.

None of the characteristics proved to be consistent with the official explanation of "military flares," and from a historical perspective, this explanation ranks alongside "swamp gas" and the

---

[17] During the Phoenix lights incident, two separate sightings occurred. The first sighting lasted one hour and fifteen minutes and the second sighting lasted thirty minutes.

"light of Venus" as a viable explanation for UFO/UAP activity which has all of the markings of a legitimate sighting[18].

Unfortunately, there is a long history of gaslighting which has been levied by government officials and representatives when it comes to the UAP/UFO phenomenon. Even with the previously discussed 1997 mass sighting in Phoenix, the immediate aftermath in response to the event by the governor at the time, J. Fife Symington was that of ridicule directed towards the multitudes of witnesses. This was accomplished by dressing up his chief of staff in an alien costume and rolling him out in front of the media as the "perpetrator" of the alleged UFO incursion. As a result, the public sentiment towards the governor negatively shifted, as the thousands of witnesses most likely felt slighted, disrespected, and gaslighted. In essence, they were being told by their governor that they didn't see what they quite obviously saw. This is par for the course when it comes to government explanations of UAP cases.

Because of this attitude, many witnesses refuse to come forward with reports. Many people (traditionally) did not want to risk their reputation, and those sentiments continue to exist today despite

---

[18] See President Gerald Ford's response to J. Allen Hynek's explanation of the March 21st, 1966 sighting by 40 eyewitnesses in Ann Arbor, Michigan which included 12 law enforcement officers: https://www.fordlibrarymuseum.gov/sites/default/files/pdf_documents/library/document/0054/4525586.pdf

the fact that great progress has been made in the past few years with respect to bringing the subject into the mainstream.

Interestingly, after the governor was out of office, several years later he did admit to also witnessing the incident and admitted that it did defy explanation. In a Fox-10 Phoenix interview on May 8th, 2014 (see link in footnotes)[19]. The former governor admitted to personally witnessing the event and stated on camera that he did believe this to be a true UFO and not a military flare or military aircraft. The governor himself has first-hand experience in these matters as a former military himself, he served in the United States Air Force. Also, as governor, he was quite familiar with the various military exercises that occur out in the deserts of the southwest United States. One might speculate that at the time of the incident, the governor may have been under political pressure to make light of the situation, again keeping in compliance with the advice put forth in the 1960 Brookings Institute publication. This approach may be the best approach from the perspective of public safety, keeping in mind that within a community, there are a significant number of people who could be adversely affected by this type of revelation.

Remember that the majority of citizens rely on the government for help and guidance during times of adversity. Thus, revelations of this nature have the

---

[19] https://youtu.be/v1Fhog5wJ7A

potential to disrupt this inherent trust and reliance upon the government to protect its citizens.

Some concluding thoughts about the UFO/UAP phenomenon involve the paucity of significant data which inevitably occurs when the government takes it upon itself to investigate reports of UFO/UAP. We have seen this in the distant past with Project Bluebook from 1947-1969, AATIP in 2007 and AARO in 2022. Historical research has confirmed that the original and longest investigation, Project Blue Book, never even received the most credible reports for a couple of different reasons[20].

First of all, witnesses were reluctant to give reports to this government agency because they were concerned about career and personal repercussions. Secondly, reports that were assigned a classified status never even breached the desk of blue book investigators on the basis of national security. Thus, if blue book set the precedent for UFO investigations, why would we think that any of the subsequent investigations (i.e., AATIP & AARO) would be any different?

Furthermore, when it comes to witnesses agreeing to give reports, one could certainly see how they would be afraid since we have all heard stories about witnesses who have lost their careers and reputations and occasionally have suffered even more ominous

[20] Lorenzen, C. E., Lorenzen, J., & Salisbury, F. B. (1967). Flying saucer occupants. In *New American Library eBooks*. (pp 149-150).

losses. Thus, most of the credible cases fall on the shoulders of civilian investigation agencies currently, MUFON, NUFORC, CUFOS, and historically, NIDS & APRO. This is not an exhaustive list, as many civilian investigation agencies have come and gone throughout history. Various reasons for this exist; however, since most rely upon volunteers, I would suspect it is difficult to recruit and retain individuals for what, in large part, is thankless and sometimes risky work. Luckily, since the UFO phenomenon has become more mainstream during the last 8 years, the risk to the reputation and credibility of witnesses and investigators is more attenuated compared to the bulk of UFO history. Hopefully, this trend will continue, and the subject will eventually be regarded as unquestionably legitimate.

Chapter 4
Crashes, Landings, & Contacts

There is a growing call by the American public for disclosure regarding the UFO topic. Most Americans today do believe in extraterrestrial life and a large portion of them also believe the government knows more than it's willing to disclose. We have had numerous credible witnesses come forward, some of whom have even testified under oath in front of Congress, most recently in both 2023 and 2024. Still, despite these affirmations, there is no tangible evidence readily available to see. Herein lies the frustration of the public, UFO researchers, and newly minted congressional officials who have recently heard recent testimony from current and former government officials, military officials, and decorated military pilots about the UFO topic.

The most compelling aspect of the recent testimony was the under-oath admission that government actors are aware of and in possession of extraterrestrial vehicles and biologics. Although hearing compelling testimony and seeing video and pictures is profound, as they say, "the proof is in the pudding." As such, I think most would agree that it's time for the secret-keepers to lay their cards on the table. Frustration among the public, members of Congress, and researchers is growing exponentially and could reach a critical mass soon, triggering a domino effect with the end result being a comprehensive uncovering, which includes hardware, biologics, and a repository of data associated with this topic. In the meantime, let us explore some of the early research regarding UFO crashes and retrievals.

The bulk of the work on flying saucer crash retrievals historically has been built off the research of Leonard Stringfield back in the 1960s and 1970s. His literary volumes go into extreme detail regarding specific cases which have been gathered from anonymous sources. This was a criticism of Mr. Stringfield at that time, being that his sources were immune to third-party verification. However, since then, various researchers have taken his initial work with several cases and expanded the field by finding credible individuals who have been willing to go on the record. Furthermore, his initial research has been augmented by field research with particular cases, such as the Aztec case.

During the month of March 1948, a flying disc was said to have crashed on the top of a lonely mesa top. It was reported to be a large metallic-like disc with a dozen deceased humanoid beings hieroglyphic writing on the inside walls of the craft was allegedly discovered. Examination of the rock formation on the Mesa where the vehicle made contact did reveal a footprint that did appear to have some markings of artificial origin, as was extensively researched by Scott and Suzanne Ramsey, who are the formative experts on this particular case. In their groundbreaking research and subsequent book, they have meticulously hunted down both ancillary eyewitnesses through both fieldwork and public records, which weave together facts that fit the puzzle and provide unconfounding evidence that is difficult to refute by even the seasoned skeptics.

Following in the footsteps of Mr. Stringfield, the Ramseys have taken his initial crash retrieval research to the next level by introducing the aspect of field research and archival research into their toolbox rather than simply relying on anonymous accounts from witnesses who cannot be verified by independent sources. However, the credit for bringing crash retrieval research into the mainstream certainly goes to Mr. Stringfield.

One of the questions surrounding crash retrievals that seems to arise is which cases are actual crashes and which cases are planned landings. Once this question is considered, this can now evolve into many speculative questions, such as how these vehicles crash since they are supposedly of an advanced technology. Critics and skeptics always point to this as if these advanced technologies and presumed beings are immune from making mistakes or having accidents. One may consider that all species are susceptible to errors and mistakes. This is seen in the animal kingdom every minute of every day. The predator and prey life of wild animals relies upon the inability of prey to be cognizant at every moment of every day, which dangers await. And, when they least suspect danger, the prey then falls victim to the prowess of the predator who lays in wait for the inevitable moment of weakness and/or inattention of their victim. Perhaps this same idea can be extrapolated to advanced beings to some extent, rendering them susceptible to environmental dangers. In the case of the Aztec, New Mexico crash, local citizens arrived first on the scene including oil workers, police officers, and a clergyman. Those individuals were to bear witness to a remarkable scene which included deceased beings and a saucer-shaped craft that was unlike any other they had ever witnessed before. Later on, the military arrived to confiscate the materials and bodies and to intimidate the witnesses into secrecy, which continued for years until many would not come forward until they were elderly and infirm.

All of the primary and secondary witnesses independently described the same key features at the site of the crash, including bodies which were the typical 3 to 4-foot tall, with larger than human-sized heads and large, widely-set eyes, and adorned with tight-fitting, "flight-suits" which almost appeared to be baked into the skin of the being. On the face, most saliently, the skin was described as a "marbled" and thickened appearance. One gets the impression that despite the fact of this phenomenal event, the local citizens were taking this in stride, as many of these witnesses had reported seeing unusual flying objects in the weeks and months preceding the actual crash. This was explicitly reported by the police officers who were on the scene of the crash. Typically, these officers routinely worked the rural stretches of highway in this area of New Mexico during the wee hours of the night and early morning, which revealed its share of anomalous activities in the skies.

One could make the assumption that as this area was home to military installations such as Los Alamos,[21] local residents were accustomed to seeing things that they assumed were part of a military project and thus became immune to any element of that which would be considered phenomenal in any other community. Furthermore, these military installations were close neighbors in the community and supported a significant part of the local economy both in a primary and secondary sense (consider ancillary businesses which support military personnel both on and off the base, including restaurants, stores, gas stations, barbers, doctor's offices, etc.) one might assume it to be very easy for military personnel to successfully install the required secrecy within the townsfolk, as many of their livelihood was dependent upon the success of the military complex. Thus, citizens would be more inclined to abide by the insisted-upon level of secrecy which was assigned in light of this phenomenal event.

One must realize that this was a much different time in history. There was no internet, social media, cell phones, etc. The social circle of most individuals

---

[21] This installation in New Mexico was a top-secret facility for the procurement of the Manhattan Project in 1942 (https://www.nps.gov/mapr/about-losalamos.htm).

was limited to immediate family and friends within the community. People's livelihoods were dependent upon this limited social circle and resources within the greater community were limited. One's reputation and "word" were taken at face value and people would do nothing to risk this. Local authorities, including police, military members, teachers, and officers of the court, were looked upon with great reverence by the average individual. Thus, the average individual would never put himself or herself at risk of diminishing their standing within the community.

It literally could mean the difference between living a comfortable life versus a life of hardship. One could see that when taken in this context, self-preservation is paramount to these civilians who happened to stumble across this crash disc. Hence, a major criticism of skeptics and critics of the UFO phenomenon is the idea that a secret this great cannot be secured for all these years. However, when viewed within the self-preservation context, it becomes easy to understand how this secret has endured the test of time. The injection of fear from the authority of these figures is enough to keep things in check and to discredit any potential threat to the liberation of the secret.

Supportive evidence of flying saucer crash reports is rich with credible witnesses. However, production of the actual hardware has been largely lacking. This is a frustration for which many researchers have spent many hours combing through witness lists and archives trying to pinpoint crash locations which subsequently lead more than a few researchers on a scavenger hunt. Ancillary artifacts have been found; however, none of which have proven beyond reasonable doubt their authenticity. Adding to this evidence is the 1994 reproduction onto 35 mm film of a manual which appears to provide instructions to crash investigation and recovery teams which was leaked to the prestigious researcher, Don Berliner[22].

This manual meticulously outlines instructions for the recovery and packaging of not only extraterrestrial hardware but also the preservation techniques expected of any recovered biological samples. This manual has been long debated by researchers, but in recent years, with the revelations offered by ex-government officials such as David Grusch and Luis Elizondo, the pendulum is certainly favored in the direction of its authenticity. The bulk of the technological period which has spanned the last 85 years has been dotted with thousands of eyewitness reports, many of which are from extremely credible individuals who would not risk their livelihood or community reputation on the basis of perpetuating a hoax. Even given the fact that most reports are explainable as mundane phenomena, still a sizable number are not explainable. Thus, given the mere bulk of official reports taken over the past 85 years, one cannot disregard all of these as frivolous,

---

[22] as cited in Friedman, 2005, *Top Secret/Majic. Operation Majestic-12 and the United States Government's UFO Cover-up*. pp, 167-184.

mistaken identities or hoaxes. Thus, at this juncture, given all of this cumulative evidence, the burden of proof is actually upon the government to disprove the existence of non-human intelligence (NHI) interacting within our reality.

Arizona is home to one of the best documented cases and was researched extensively by the investigator, Harry Drew. Reports from credible witnesses confirm not so much a crash but an unconventional landing of a saucer on May 21st, 1953. Legend has it that this occurred approximately eight miles northeast of the airport in Kingman and the craft was studied by a team of 40 engineers and scientists who were sworn to eternal secrecy regarding the matter.[23] The highway was too narrow to pass through the section which traversed over the Colorado River at the time. Thus, the saucer was loaded onto a barge which crossed the river adjacent to the Hoover Dam.

Legend has it that a portion of the saucer impacted the dam as it was making its way across the river, most recently reported by researcher Michael Scratt on a 2019 Richard Dolan podcast. When discovered, the craft was largely undamaged despite estimated contact with the ground at a sizable velocity. Witness accounts consistently reported this craft as being approximately 30 feet in diameter,

---

[23] https://www.azcentral.com/story/news/local/arizona-contributor/2016/11/15/did-ufo-crash-kingman-1953/9 3828300/

saucer-shaped, and partially buried in the sand, which presumably occurred with the impact. Extraterrestrial Biological Entities (EBEs), estimated to be approximately 3.5 ft tall with large craniums and large eyes, were reportedly found deceased and wearing tight-fitting, metallic-like flight suits. Interestingly, there were other reports that some EBEs actually survived this impact and were recovered by authorities as well. Along with this are reports that the Kingman saucer was actually given to the United States as a "gift" by these EBEs. Artifacts were loaded onto heavy equipment and transported via both roadways and waterways to the final destination at top-secret facilities in Nevada and Ohio. Could these humanoid explorers be responsible for planting the seeds of our humanity in an effort of intelligent design? Perhaps there have been countless intermediaries between our precursors and our current state of being. Maybe the Greek mythology of Centaurs is just one of those early intermediate steps with their torsos of men and their caudal regions of horses[24].

One could only speculate; however, it is important to have an open mind to be truly ingrained within the scientific method. The humanoids that have been described in the majority of these credible crash reports fit the criteria which are outlined in the operations manual which was craftily reproduced and outlined in Stanton Freidman's original 1996 Top

---

[24] World Mythology. The Illustrated Guide (1996). p. 164.

Secret/Majic publication and are referred to as Extraterrestrial Biological Entities or EBE for short. However, sources differ substantially with regard to height with beings described as anywhere from 3.5 ft tall to 5 ft tall. One may speculate that depending upon the place of origin, these EBEs may pose the same variability in height as we find with various ethnicities on earth.

The Rendlesham forest incident, although not exactly a crash retrieval, is an incident worthy of study insomuch as it has all of the markings of a genuine contact case. Here, military personnel stationed at RAF Woodbridge in the U.K. (circa 1980) reported contact with an object in the forest and some of the soldiers involved in subsequent years have developed chronic physical ailments which are believed to be directly linked to the incident. In fact, some of these physical effects have only recently been settled by the U.S. military in terms of granting disability benefits for one of the soldiers involved. However, it is important to mention that to the best of our knowledge, no admission of the exact etiology of the physical ailments has been provided by the government publicly.

Also involved in this incident were effects on the vegetation within the forest, including burn marks on the trees. Researchers have concluded in the years following the incident that both the physical effects and arboreal effects are most likely the result of electromagnetic frequencies which were being produced by the object in question. Skeptics of this incident have blamed this sighting on a beam from a nearby lighthouse, a conclusion which quite clearly lacks any serious investigation of the well-documented facts of the case. UAP researchers have skillfully debunked these claims using supportive evidence and eyewitness accounts from those directly involved in the incident and reported in declassified government documents.

Some conclusions can be inferred from the cumulative documentation regarding crashes, landings, and contacts. Most importantly, there has been no evidence in the historical record which demonstrates subversive intent from these crafts. Thus, despite the reports most recently cited in the mainstream media regarding the potential “threat” posed by UAPs/UFOs/Saucers, there is no evidence which demonstrates any aggressive action of a primary nature. This does not mean there hasn't been a loss of human life. On the contrary, there have been instances where acts of aggression initiated by military aircraft have unfortunately led to the untimely demise of airmen, presumably because of defensive actions taken by the craft.

Additionally, as we will explore in another chapter, there is some evidence of behavior which may be considered aggressive towards cattle. However, as we will see, there is a large question as to whether or not these cattle abductions are actually hostile in nature or, more than likely, the cattle are euthanized in a humane fashion. As far as crash retrievals are concerned, there is evidence that at least a portion of these incidents are due to deliberate acts caused by nations. There is no consensus, however, as to how these crafts are brought down, but some speculation exists that it could be due to deliberate radar interference of some sort. Moreover, there is speculation that once live beings are taken into government custody, their treatment might be less than fruitful, as rendered by authorities with the goal of garnering information from said visitors.

Unfortunately, it seems that most tales of saucer crashes do not end well for the pilots of the craft. They are either DOA or taken into custody by officials who may treat them less than appropriate during their tenure, only to end in their premature demise. Except for the promising tale of humanoid aliens in the 1960s who were quite helpful to our government officials and supposedly even had an exchange program where a baker's dozen U.S. military personnel spent 13 years on their distant exoplanet (SERPO), most of the crash retrieval stories are quite ominous for the beings who crash-landed on our rocky world. In contrast, the human visitors to planet SERPO were welcomed with open arms by the humanoid race of extraterrestrial biological entities who inhabited this planet which is 39 light years away and basks in the light of a binary star system[25].

---

[25] For detailed information, read the book "Secret Journey to Planet SERPO" by Len Kasten.

The big Kahuna of crash retrievals is indeed the Roswell incident. To spend any amount of time on this would certainly not do justice to the impeccable research done by Stanton Friedman in his 1992 edition of "Crash At Corona." As my homage to Mr. Friedman, you have probably noticed that I favor the term, 'flying saucer" because this was Stan's salient point throughout his tenure; that he did not care about UFOs, he cared about flying saucers and another term he was known to coin, "Earth Excursion Modules." The former and latter terms give much more character to this subject and I will therefore borrow them in homage to Mr. Friedman. I do want to make a few points about this pivotal event in the crash retrieval history however, for academic purposes.

When this crashed flying saucer was discovered on a ranch outside of Roswell, New Mexico, on July 8th, 1947, this was the first time in modern history where you had an almost immediate disclosure come from the military and leaked to newspapers and radio stations. There was no washing of the information. It was leaked "as is" to an unsuspecting public. It seems that once the military knew the value of what they had discovered, it was too late because the information was already in the hands of the American public. Their only recourse was to retract the information, but as the saying goes, "the horse was already out of the barn," and subsequently, this fueled the beginning of distrust in government which in the subsequent 77 years has only evolved into a roaring fire.

Furthermore, to make matters even worse, in subsequent years, the government has attempted on three separate occasions to provide alternate “truths” about what really happened that day. The first excuse was that the crashed saucer was actually a weather balloon. The second excuse given in 1994 was that it was spy equipment used to monitor Russian nuclear tests. The most recent excuse provided by the government in 1997 was that the alien bodies reported in the original incident report of the Roswell incident were nothing but crash-test dummies released from the upper atmosphere. The only problem with these subsequent excuses was that the time frame did not match the original incident and the reports directly conflicted with reports of eyewitnesses on the scene. Common sense tells us that when a person or government entity changes its official explanation of a documented event, there's most likely a significant element of dishonesty occurring, even when given the benefit of the doubt.

Though this does not “prove” that the Roswell incident was indeed an alien crashed vessel, it does cause much suspicion among researchers who continue to meticulously study this incident within the realm of declassified documents which continue to be released through FOIA requests. Credible, independent witnesses have confirmed the presence of humanoid bodies and hardware recovered from this crash. Moreover, a common theme in their account of the incident was being instructed by government/military officials to basically either forget what they had seen or that what they had seen were nothing but mundane artifacts of unexotic origin.

Lastly, some of the witnesses claimed that these artifacts and bodies had eventually meandered into the custody of military bases in the southwest and midwest United States[26]. Claims from retired military personnel in the many years subsequent to 1947 up until the modern day have corroborated many of these original eyewitness claims. This begs the question, are all of these witnesses lying or delusional? Logic would dispute any claims of such, considering the plethora of historical evidence garnered over the past 77 years. Furthermore, a sizable sample of the American public believes that some degree of a cover-up has indeed transpired. There's certainly power in numbers in this regard.

Another case which happened many years before Roswell, back in 1941, has been quite elusive in terms of hard evidence but is still worth mentioning in terms of its historical relevance is the supposed crashed saucer incident in Cape Girardeau, Missouri. The lone witness was a clergyman who was tasked with giving the last rights to the three dead and/or dying humanoid occupants before the military again whisked away the hardware and bodies to unknown destinations. The little-known details include the stereotypical 3-4 foot tall humanoid aliens with a wide set, large eyes, a disc-shaped craft with hieroglyphic

---

[26] e.g., Area 51, Los Alamos, Wright Field (Dayton, Ohio).

type writings on its skin surface, and a bland control panel with continuous and contoured surfaces. Modern descriptions of this craft would be consistent with that of a 3D printed construction without any nuts, bolts, or fasteners and almost akin to being injection molded into one large continuous piece. This type of description of craft is quite common throughout the tenure of flying saucer witness reports which have spanned the modern 80-plus years. Unfortunately, the subsequent 83 years after this incident have not resulted in verification from independent sources and have rendered a paucity of documentation, in stark contrast to the Roswell incident and the Aztec case.

In September of 1994, over 60 children of middle school age witnessed a landed flying saucer with its occupants walking around the vehicle in Zimbabwe. This magnificent event happened while the students were at the playground during recess but their teachers were having a meeting inside the school. This was a mass sighting where the majority of the students reported the same or very similar events, and now, almost 30 years later, these students continue to tell the same tale. Over the years, the incident has been labeled a hoax by some skeptics, while others have called it a mass hallucination. However, the teachers of the students found no evidence that students made up the story and examinations conducted by Harvard Psychiatrist Dr. John Mack found no evidence of mental abnormalities to explain the sighting.

A similar sighting took place in 1966 which involved 200 students and teachers at a high school who witnessed the landing of a silver disc UAP near their school in Melbourne, Australia. Both of these cases are particularly credible because of the fact that they were mass sightings of mostly school-aged children. Thus, one can conclude that the chances for ulterior motives, such as financial gain or notoriety, are quite low in comparison to adults, whose chances for ulterior motives would be much higher. Furthermore, children are inherently truthful when it involves significant and/or traumatic events due to the fact that they must rely on their adult caregivers for their needs and well-being. Thus, in response to authority figures, they are more apt to tell the truth in comparison to adults, who may withhold pertinent information due to fears of being viewed in a negative light. In a sense, one might conclude that children may be even more credible than adults as witnesses to events such as these, as their motives are generally pure and unbiased.

Another controversial incident involved sightings which occurred in Gulf Breeze, Florida, in November of 1987 and involved a local contractor, Ed Walters, who managed to snap several polaroid pictures of a saucer-shaped object hovering approximately 200 feet above the ground just outside of his home (see figure-1). Several other residents in the area reported seeing similar craft hovering. The photos were studied by Physicist Bruce Maccabee Ph.D who rendered an opinion that they were authentic. Skeptics proclaimed the Gulf Breeze sightings to be a hoax. This incident is debated even today about its authenticity and has its share of supporters and skeptics. The Gulf Breeze incident garnered national attention and appeared on a popular prime-time television show, Unsolved Mysteries. Examination of the photo, in this author's opinion, definitely demonstrates authentic attributes as one would expect with a craft that operates with a presumed antigravity propulsion system as it reacts with our gaseous atmosphere. As a result, it created a luminescent glow and, as such, appears to exist both within the atmosphere and outside our atmosphere simultaneously. Thus, it appears this craft is both solid and translucent in the photo.

There exists a plethora of substantiated evidence which has been independently verified by professional researchers through solid investigative means. Furthermore, the credentials of these researchers are fully vetted and span a multitude of academically grounded and well-respected disciplines such as engineering, microbiology, physics, health sciences, materials science, astronomy, astrophysics, and aviation professionals (just to name a few). This begs the question of disclosure and whether or not it means anything at this point. Collectively, those of us who have invested years of dedication and energy to this subject, certainly do not need this "seal of approval" as it's surely never to come. At least not in a fashion which would afford this discipline the respect that it deserves and for which we, its stewards, are responsible for both its preservation and advancement.

Figure-1, Source:
https://unsolvedmysteries.fandom.com/wiki/Gulf_Breeze_UFO

References:

https://www.pnj.com/story/news/2017/12/10/gulf-breeze-ufo-sightings-30-years-public-divided/915397001/

Chapter 5
Crop Formations

This chapter focuses on crop formations, as there seems to be a natural relationship between the UFO/UAP phenomenon and the formations to some unknown extent. Most of this presumed relationship, I will admit, is based upon many years of eyewitness accounts, which seem to include observations of flying and/or floating objects seen within the vicinity or directly above crops of corn and wheat for which crop formations ultimately seem to manifest. Although an entire book can be written on this topic alone, I felt it necessary to include some of the basic information within this book in order to put some context into the UFO/UAP topic.

As we will see, there have been formations that seem to sometimes answer humanity's basic questions about the nature, purpose, and intent of non-human intelligence. For over 300 years, there have been documented observations of formations that have sporadically and mysteriously appeared within the fields of farmer's crops such as wheat and corn. In a significant number of these cases, there is no logical explanation of the source of these formations. In fact, some of them happen instantaneously without any prodromal weather or environmental changes, while others are sometimes accompanied by eyewitness reports of balls of light and/or orbs which seem to mysteriously float about the field, sometimes smooth and drifty and other times moving erratically prior to the production of these crop formations and designs which are typically geometrically intact. It is important to note these formations include an array of different shapes and sizes. Thus, we will refer to them as formations in exchange for the traditional nomenclature, "crop circles."

An early 20th century case occurred in 1927 under the shadow of the Cissbury Ring in Worthing, England. Roger Seer observed a geometric formation of swirled corn and brought his compass into the center of the design. He noticed the needle of his compass begin to dance around and lose directional focus when within the formation, but when he stepped out of the formation, the compass regained its directional abilities. Furthermore, when within the formation, he experienced a burning smell, a strange prickling sensation in his feet, and a headache developing. Moreover, a nearby barn dog refused to enter the formation as if some mysterious force was preventing it from doing so.

Another incident took place in south Holland which involved a formation that suddenly appeared after a magnificently illuminated bright, "purpleish" suddenly appeared and was described as being in the shape of a football. It was observed to hover shallowly over a field of crops. It then morphed into a disc and was described as giving off a type of energy which disturbed the atmosphere over the surface of the field. This appearance presumably is consistent with an EM energy burst. Subsequently, this burst of energy caused these crops to flatten into a perfectly circumscribed disc-shaped formation within the field. The type of crop in this particular field is unknown from the context of the case. However, from most reports, one could assume either corn or wheat was the affected crop.

Southern England has a significant amount of crop formations, more than two-thirds of which occur in Wiltshire, Hampshire, Exeter, West Sussex, and Devon[27]. In fact, in the recorded year 2002, 84 out of 96 formations were located in the southern locales of England. Records of these crop formation manifestations have been occurring in this part of England as far back as the seventeenth century. In fact, in 1678, a pamphlet was released entitled The Mowing-Devil or Strange News out of Hartford-Shire, explaining how an unexplained fire illuminated a farmer's corn the night after he refused to employ the neighborhood mower. The next morning he found areas of flattened corn that he thought too perfect for man to have created. Many more examples of fairy rings, crop formations and associated phenomena pepper the pages of folk history from the beginning of memorialized time up to the more modern mainstream press[28] era. One must keep in mind that in the modern day, many of these formations are

---

[27] Spatial distribution of England's crop circles: Using GIS to investigate a geo-spatial mystery Jeremy Northcote Edith Cowan University, Australia Email: j.northcote@ecu.edu.au
[28] Wiltshire Times in England on Aug 15, 198

indeed man-made for reasons that largely have to do with enriching the tourist economy of the region which has built quite the reputation for these crop formations over the past century. However, the etiology of a significant number of formations remains (estimated < 20%) unexplained. Unfortunately, the crop formation phenomenon has received a paucity of research in the many years which have passed since it was first reported in mainstream media.

The original and most notable research came from a Michigan biophysicist, Dr. William Levengood, PhD, who reported the most compelling evidence regarding plants that were involved in unexplained and spurious crop formation formations[29]. Dr. Levengood ran experiments on several plants taken from both inside and outside of crop formations. It is important to note that Dr. Levengood's reputation within the scientific community was well vetted, as he has published his research in various scientific journals, including the journal, "Nature." Much of the evidence is centered around the effects of some unknown force or energy upon the crop nodes. For those who do not have an agricultural background, a crop node is a joint in a wheat stem which appears above the soil when the plant reaches a particular stage of growth. The purpose of these nodes is for the extension of plant growth, when various shoots grow and procure leaves. Dr. Levengood studied 250 plants from within formations, which he compared to a

---

[29] https://www.bltresearch.com/print/pplantab1.html

control group of plants from outside the crop formations, and noted significant differences between the formation group and control group[30]. These differences included the hypertrophy of the cell wall in crop formation plants particularly in the region of pits in bract tissue, which is the region of the plant where a thin membrane exists which surrounds the seed head. This is responsible for being a conduit through which nutritional matter is provided to the growing seedling. Hence, this cell wall hypertrophy was not present in the control group crops.

Lateral and longitudinal expansion of plant stem nodes was also apparent in crop formation plants and was absent in controls. Additionally, the formation crop nodes all displayed evidence of bending and stretching at various node locations within the crop. Presumably, Dr. Levengood accounts for this phenomenon through some speculated type of thermal heating which occurs very quickly. This bending was typically shown to be between 45-90 degrees in the samples studied by Dr. Levengood and his team. These researchers took considerable caution not to misidentify aberrant node bending with natural etiologies of node bending, which include gravitropism and phototropism. The former is the crop's ability to follow the earth's gravitational field, and the latter is the crop's ability to attach itself to sunlight. Some other observations seen within the

[30] All of the differences between crop plants and controls were noted to have a p-value of ≤ 0.05

sample crop studied by Dr. Levengood's team were expulsion cavities, upregulation and downregulation of subsequent growth rates on both seedlings and the plant structure itself in mature crops. Hence, this growth regulation appears to vary with the maturity level of the crop, and as a result, some seeds were malformed and displayed stunted growth, and germination effects.

Rigorous study and application of the scientific method resulted in conclusions by Levengood and his team which adhered to the Beer-Lambert principle, which Levengood denotes as a well-known principle in Physics and is explained as the absorption of electromagnetic (EM) energy by matter. In these sample crops, the observed node effects were hypothesized to have occurred through the application of a focused EM pulse of energy. One of the most macroscopically salient features of the genuine crop formations, which is not appreciated through Levengood's experiments, is the fact that field investigators have noted plainly evident differences between hoaxed formations and genuine formations when viewing the patterns from ground level. In the genuine formations, the crop stems are flattened in a manner that demonstrates patterns in both height as the formation traverses both inward and outward from the formation. Moreover, the flattening is performed in a manner that demonstrates a carefully constructed "weaving" pattern which human creators are simply not capable of achieving through traditional methods of creating man-made crops which involves manually cracking the stems of each individual plant within the formation.

Let's speculate for a moment regarding the etiology of the formations. Suppose, yes, it is true that in accordance with the work of Dr. Levengood, the effects on the crops are that of some focused, electromagnetic (EM) pulse which creates changes within the stem of the plants which originate from the nodal points of attachment. Dr. Levengood reached the EM conclusion by virtue of the fact that he was able to reproduce the node changes in healthy crops by exposing them to microwave energy[31]. This would point towards an artificial energy source as the etiology of the genuine formation crops versus a natural source. As such, Levengood attested that for humans to accomplish such a feat, highly specialized equipment would be needed, and this he felt was highly improbable. This author would also concur with Dr. Levengood's assessment. As a result, the plant is bent in a fashion or to a certain degree which is structured in such a way that when within the formation of other crops, a geometrical configuration is produced, which in effect creates some meaningful pattern as interpreted by human beings. One could thus speculate a "natural" etiology for this or, possibly, an "intelligent" producer is responsible for this. Perhaps the reports of orbs and flying objects within the vicinity of these crop formations are

---

[31] The Chronicle. Duke University. Durham, North Carolina. 10/01/1991. Volume 87, #23.

credible, and perhaps there is a link between these two observations.

Moreover, in support of this is the fact that a common theme with UAP/UFO contact cases is the multiple reported occurrences of what is described as electromagnetic pulse energy which has disabled the electrical systems of motor vehicles, aircraft, and even nuclear missile launch systems. How could scientists design a study that can observe and measure this phenomenon in a quantitative fashion? Since scant evidence has been recorded and imaged when these events occur, perhaps a study proposal might involve establishing video surveillance technology with the ability to record infrared and randomly assigning them to geographical areas where there has been a dense historical accumulation of formation events.

One geographical area to consider such an experiment may be the area of Wiltshire, England which has the highest number of reported formations compared to other counties in England. Wiltshire boasts a whopping 380 formations since 2005, which typically occur overnight in fields of wheat and corn. As with the majority of topics that are considered to be outside the realm of mainstream science, funding for this type of endeavor is limited to small private sources and individuals. Typically, no large government-sourced grants are allotted for such projects which, unfortunately, have been labeled by legacy scientists as "pseudoscience." Hence, the research performed by W.C. Levengood, despite showing significant results, was small and unreplicated.

Furthermore, many legacy scientists found it quite easy to criticize the methods and outcomes achieved by W.C Levengood since his research lacked the large scale, handsomely funded, and legacy journal sponsorship enjoyed by scientists who lend their research to more "acceptable" topics. In fact, this pervasive criticism by legacy scientists is often enough to be a career destroyer for any serious scientists hoping to expand their knowledge within this subject. This common trend among topics of this nature can be viewed throughout modern history and is part of a problem that has been recently highlighted by Professor(s) Avi Loeb of Harvard University and Garry Nolan of Stanford University, who both have spoken publicly about the attitude which historically plagues legacy scientists to remain closed-minded to topics of interest which are outside that of "accepted" possibilities.

Dr. Loeb often instills the notion that these mainstream scientists serve as gatekeepers. By maintaining a closed-minded approach, they work against the very principles of science—discovery, openness to new ideas, and, by extension, new opportunities. Certainly, science is not intended to fit into any predetermined narrative put forth by a small number of individuals who are trying to achieve or preserve a particular agenda. Professor Loeb and, before him Carl Sagan, have both used the analogy that the best scientists are the children because their questions come from a place of innocence and pure curiosity. Hence, their immaturity, in a way, protects them from the hidden agendas which adults are often controlled by. Unfortunately, this does not hold true once adulthood is reached and reputations pave the way to career advancement opportunities which most deem too delicate to risk on subjects that are viewed as fringe.

One of the problems that plague the field of Cereology is the constant criticism of skeptics who are persistent in blaming all occurrences of crop formations on hoaxers. In the common literature, you will find the names Doug Bower and Dave Chorley, who claimed to have created over 200 crop formations during the wee hours of the morning which typically occurred after a night out at their local pub in Southampton, England. Once this story was published in local newspapers, it caught on worldwide, and the skeptics immediately latched onto this story for which there is actually scant evidence to substantiate their claims. First of all, a significant number of these crop formations were found to have such complicated geometric patterns that it would be nearly impossible to coordinate efforts to create these patterns in the still of the night, oftentimes without any witnesses and without any evidence that human beings had stepped foot into the fields. In fact, upon inspection, oftentimes, there were no tracks leading to or from the fields and no evidence that any human being was there to begin with. No garbage was left behind, no evidence that anyone had taken any bathroom breaks. This would be virtually impossible to accomplish based upon the complexity of these formations and the fact that it would take several hours literally to create these spectacles and without any durable lighting except for handheld flashlights, "supposedly."

Pure common sense would indicate that in order to accomplish this, humans would need proper equipment facilities for taking breaks using the bathroom, and what about the potential injuries that would most definitely occur when trying to accomplish this in the dead of night. Surely, someone would cut themselves, sprain, fracture an ankle, throw out a back, etc. However, there have never been any reports of incidents such as these, especially in crop formations, which demonstrate all of the signs of being legitimate (i.e., recall the novel properties of William Levengood's experiments). Don't confuse my words, for indeed, yes, there are a significant number of formations which are man-made. But typically, the construction of manmade formations is not a hidden feat. These formations typically draw much attention from bystanders and the creators must surely procure permission from the owners of the crop fields prior to trespassing onto their field.

Moreover, man-made crops typically have broken crops instead of bent and stretched crops due to the fact that man-made crops are created in such a way that tools are used to crack and subsequently flatten the crop. The tool typically consists of a wooden plank which is attached to a set of ropes and is referred to as a "stomper." This creates much botanical trauma which is quite easy to identify as separate from the likely EM process which creates genuine formations. Although the designs of which humans are capable have become more intricate in modern times due to the advent of GPS and computer-aided technologies, one still cannot get around the fact that human creators will cause significant trauma to the crop, unlike the process by which the unknown creators utilize.

Another feature of some crop formations is the fact that when a compass is brought within or near some of the formations which are thought to be genuine, the needle of the compass often acts erratically and even sometimes points in an unexpected direction. Although these effects are anecdotal, it is reported often enough that pilots often avoid areas known to have crop formations due to the possibility of unexpected effects on navigation equipment.

Moreover, some researchers have proposed a relationship between the geometric ratios of some crop circles to that of musical harmony which is reflected as to match various octaves of pitch upon the diatonic scale which includes C, D, E, F, G, A, & B (Hawkins, as cited in, Howe, 1997). Interestingly, skeptics often point to the fact that there are no large, traditional scientific studies on this phenomenon in the sense that we expect within academic journals, such as double-blinded, placebo-controlled, quantitative analysis. However, when it involves qualitative research and observation, the field of Cereology is rich with data, much like that of the field of Ufology. Mainstream scientists often discount eyewitness testimony and observation in favor of expensive quantitative analysis. However, in the end, it is the human experience that accounts for the context by which the world is interpreted. Hence, scientific discovery is meaningless unless there is a human being there to witness it and filter it through the physical (cerebral) and cognitive pathways that interact to provide us with a snapshot of the world.

Furthermore, this is only part of the picture since we have not considered the effect of consciousness and the ways it could interact with this three-dimensional world. Perhaps human consciousness is the key that activates the fourth dimension and beyond to guide us into a cosmic architecture that otherwise is not accessible. Again, as scientists, we should keep an open mind with regard to these possibilities.

One of the most fascinating formations occurred in Hampshire, UK, in 2001, termed the Arecibo answer (see Figure 3 below). In 1974, Frank Drake and Carl Sagan, in conjunction with the SETI Institute, sent a radio message into space consisting of 1679 BITS of information embedded within a radio signal of 20 trillion watts out into the galaxy. The information encoded within the message gave specific details about humanity (our DNA), our planetary system, and the biochemical nature of earth life forms. The Arecibo answer was the purported response from an unknown alien intelligence to the original message which took almost 27 years to receive. Interestingly, the original message was directed to the globular cluster M13 which is 25 light years away. Despite this, the SETI Institute was quick to conclude that the Arecibo response was a hoax. It is interesting to note, however, that the response message included such details of the alien civilization which included their biological construction of silicon and their DNA containing an extra strand. Furthermore, their physical characteristics revealed that of a large head and small body (4 feet tall). Their population of just over 21 billion is claimed to be dispersed upon three separate planets within their solar system. Unfortunately, there is no published information about whether or not the crops from the Arecibo formation demonstrated any of the properties of genuine formations identified by W.C. Levengood in his original research.

In conclusion, we are left with a situation where, yes, crop formations are a real occurrence, and it is safe to conclude that not all crop circles are man-made. However, we cannot say with 100% certainty what forces are responsible for the creation of genuine circles. There is some robust evidence that demonstrates EM energy is responsible for the changes observed in genuine formations, such as the bending and stretching of crop nodes and perhaps the intricate weaving patterns that are observed in genuine formations. Despite the fact that we cannot confirm an extraterrestrial etiology for these formations, it may be possible to consider some influence by non-human intelligence. What is required is brave scientists who are not afraid to ask these questions and perhaps establish quantitative analysis to build upon the original work that was started by Dr. Levengood.

References:

Howe, L. M. (1997). Glimpses of other realities. LMH Productions.

Chapter 6
Origins

Keep in mind that the terms host star, star, sol, and sun are used interchangeably as you read through this book. Even though there are obvious differences with regard to classes of stars, for simplicity's sake, we can assume these terms collectively reference any of the billions upon trillions of heavenly bodies within our universe, which are actively completing fusion reactions by which hydrogen is benign converted to helium. This is the primary method of energy production during the main sequence life stage of a star.

An August 6th 2024 article in Space.com[32] outlined ten possible earth-like exoplanets within our galaxy which have been identified by space telescopes such as Hubble, Keppler, and James Webb over the past several decades. Although much is still unknown about the atmospheres of these exoplanets, scientists are acquiring new techniques each day to determine whether they encompass the recipe needed to sustain life as we know it. That is, carbon-based life forms. However, this does not rule out the possibility that other chemically based life forms exist, such as silicon-based life forms. It is important to note that what we consider as earth-like from a human perspective involves particular characteristics. First,

---

[32] https://www.space.com/30172-six-most-earth-like-alien-planets.html

we make the assumption that earth-like planets are rocky planets and are roughly the same size as our earth. Also, we make the assumption that earth-like planets orbit their star in an elliptical pattern, and most likely, they also rotate about an inherent axis similar to our north-south axis rotation. Another assumption that can be made is the presence of moons such as our own, which helps to create the tides of the bodies of water which presumably exist upon the surface of the planets. As such, we assume that earth-like planets have water on the surface which is made possible by the fact that they are both close enough and far enough from their host star so that it is possible for water to exist and for temperatures to be without inhabitable extremes. We call this zone of habitability the "Goldilocks Zone."

Several candidates have been identified as the "Kepler" planets as they were first identified by the Kepler space telescope. These include Kepler 667Cc, 22b, 69c, 62f, 186f, 442b, 452b, & 1649c. Also identified were Proxima Centauri b, & TRAPPIST 1e. Another possible candidate star which potentially may have a habitable exoplanet is Tau Ceti. This star is considered a G-class star, just like our sun. In cosmic terms, it is considered relatively close in distance at 12 light years and is within our galaxy. The binary star system Zeta Reticuli is situated 39 light years from our solar system and is within the Milky Way galaxy. Although there is no reported astronomical evidence to date of exoplanets in this system, this star system has a rich history within the UFO lore as being home to alien races of intelligent beings. Interestingly, the two stars within the Zeta Reticuli system are G2 & G3 class stars and thus are similar to our sun. The renowned researcher Stanton Freidman has written about these stars as being included within a group of stars within 100 light years from earth which would be likely candidates as the origin of various species of intelligent beings who have visited our planet.

Methods of direct imaging of exoplanets are still in their infancy and continue with difficulties when it involves visualizing earth-like planets; in particular, because of their small size, this host star typically overpowers the ability to image it. The most common method and that which has been able to detect the most exoplanets is the transit method which looks for changes in the light intensity of the host star which occurs when exoplanets pass in front of it. This method has been able to identify 4,309 exoplanets, according to NASA[33]. Additional methods include Radial Velocity which has been able to identify a slight "wobble" of an exoplanet's host star which occurs when the planet pulls on the host star as it completes its orbit. This is caused by the fact that the host star's gravitational field interacts with that of the planet as it completes its orbit around the star. This method has identified 1,094 planets, according to NASA (see previous link in the footnotes). Other less common

---

[33] https://exoplanets.nasa.gov/alien-worlds/ways-to-find-a-planet/?intent=021#/1

methods include gravitational microlensing and astrometry.

To have a rudimentary level of understanding of the physics involved in these various techniques, one should familiarize oneself with historical aspects of astronomy, particularly Kepler's three laws of motion.

We have already touched upon the first law which states that planets orbit their host star in an elliptical fashion whereby our star (sol) is situated at one fixed focus of the ellipse. Kepler's second law basically describes a path by which planets gain speed as they become near to the host star and decrease in speed as the star moves away from the sun. Kepler's third law deals with the orbital speeds and effects upon orbital time relative to distance from the star. This law, for instance, explains the fact that it takes the earth 365 days to orbit the sun versus a planet such as Jupiter which is much further from the sun and thus takes about 12 years to make a complete orbit around the sun. These laws of planetary motion are true for all solar systems, presumably in the universe, as they all operate upon the basis of gravitational interactions among a host star and its orbiting planets.

Our sun is currently within its main sequence stage which lasts a total of 8 billion years. Its current age is 4.5 billion years which is more than halfway through its lifespan. Thus, another assumption is that solar systems that have the potential for intelligent life, albeit assuming that the intelligent life evolved within that system, must be roughly at least the same age if one is to assume that these life forms are at the same level of intelligence as we are. However, multiple assumptions must be true for this to be certain. This would include an assumption of carbon-based lifeforms on a similar planet with similar conditions which have evolved smoothly without any catastrophic event to delay the evolutionary trajectory or cause the trajectory to cease. Another possibility is that intelligent life forms that evolved on exoplanets which orbit red dwarf stars could be much older and, thus, more advanced than us. The characteristics of red dwarf stars are that they are the most abundant stars in our galaxy and universe and their lifespan is from tens of billions of years to trillions of years. Red dwarf stars comprise about 75% of the stars within our galaxy compared to about 8% of sun-like (G-class) stars. Their mass is typically only a fraction of our sun's mass (7.5% to 50%), and they burn at a lower temperature (3500 degrees celsius in comparison to our sun which burns at 5600 degrees celsius). Hence, there exists speculation that old civilizations may have evolved upon rocky exoplanets that potentially orbit these low-mass red dwarf stars, many of which are binary and even triple star systems.

A prime example of a red dwarf system which has been covered extensively in science news over the past several years is the TRAPPIST-1 system which contains seven potentially rocky planets which orbit a red dwarf star within our galaxy, approximately forty light years away from our solar system. Three or four of these exoplanets have evidence to support their existence within the habitable zone (goldilocks zone) of their host star which is an M-dwarf star currently about 7.6 billion years old[34] which is almost three billion years older than our sun. This system has been the focus of great speculation and study most recently by the James Webb space telescope (JWST), which is currently trying to determine whether or not any of the planets in this system contain thick enough atmospheres that would allow for life to exist. Although no definitive conclusion has been reached regarding all of the planets in this system, some recent evidence has been discovered by the JWST which shows that the innermost two planets within this system most likely do not have any appreciable atmospheres. However, this would not be unlike our own system, whereby Mercury is void of an atmosphere[35].

Another obstacle with red dwarf solar systems is the fact that their small size and low temperatures make it necessary for any orbiting exoplanets to oftentimes be much closer in proximity to the star, which creates a situation whereby the planet is tidally locked with the star. This creates a situation whereby one hemisphere of the planet is in permanent daylight while the opposite hemisphere is in constant darkness. This combined with the fact that red dwarf systems are prone to frequent solar flares, may be obstacles for life to exist. However, there may be

---

[34] https://www.space.com/35806-trappist-1-facts.html
[35] https://phys.org/news/2024-02-atmospheres-trappist.html

mechanisms by which stable environments are created upon these worlds such that life in particular environments may flourish. Indeed, refinement of techniques for examining these worlds is necessary before any definitive conclusions can be drawn. Assuming that our galaxy and, by extension, our universe does contain life forms, we must turn our attention to the Fermi paradox, which is mentioned quite commonly in the media.

The Fermi paradox, named after the Italian physicist Enrico Fermi, once questioned why we have not made contact with any of these intelligent life forms. The answer to this question reveals multiple possibilities, including the fact that we have indeed made contact and the evidence of this contact is deliberately hidden in most cases. This would be especially salient when we discuss reports of crash retrievals in chapter four. Another possibility is that a number of these earth-like exoplanets have a significant gravitational strength which precludes any intelligent species inhabiting the surface of such planets from achieving escape velocity with their current level of technological prowess. One can speculate that human-like civilizations exist in all spectrums of advancement from preindustrial to postindustrial and everything in between. Perhaps they also look up at the stars in wonder and dream of what lies far beyond their reach. The possibilities are endless in this sense and vary depending on atmospheric conditions and gravity.

Perhaps some planets with stronger than earth gravity have beings that are predominantly four-legged or maybe even three-legged with thick, strong musculoskeletal structures and robust cardiovascular power, whereas planets with thinner atmospheres and weaker gravity may have smaller, thinner bodies with less cardiovascular power. Perhaps some planets have beings that utilize silicon as the base structure of their biochemistry. Imagine the varied modes of communication such as telepathic, tactile, or perhaps vibrations are the main methods of communication. Furthermore, maybe alien communication does not have a beginning and an end, and perhaps written forms of communication can provide glimpses into the future. A multitude of possibilities exist when one considers the possible degree variations that are possible when it involves brain/cognitive function.

Some civilizations that we assume exist in our galaxy and in other galaxies have indeed reached the level of technology whereby one of their methods of mass communication is that of propagation of radio waves. The SETI institute (Search for Extraterrestrial Intelligence) is an organization which has been searching for radio signals from advanced technological civilizations for the past 40 years. To accomplish this, they have established arrays of radio telescopes around the world which continuously scan for narrow-band radio signals around 1420 megahertz because this is the most likely frequency by which alien civilizations would broadcast, as it occurs naturally in the universe. This frequency is a prominent, narrow-band signal which is dispersed throughout the galaxy (and, by extrapolation, the universe) and, as such, is agreed upon by SETI scientists as the most likely frequency by which alien civilizations around our level of technological prowess are using to communicate on a mass scale.

The 1420 megahertz frequency is the natural emission frequency of hydrogen atoms, which is the most common element in the universe. The existence of civilizations that use radio signals to communicate does not preclude the existence of other forms of communication which are beyond our level of technological achievement. On August 15th, 1977, the goal of the SETI institute may have come to fruition with the discovery of a narrow band signal near the 1420 MHz hydrogen line recorded by the Big Ear radio telescope at Ohio State University. Although the discovery of this signal certainly checked all of the boxes consistent with that of an alien signal, SETI scientists have been subsequently unsuccessful with regard to the redetection of the signal. However, according to SETI.org, a candidate star by the name of 2MASS 19281982-2640123, located 1788 light years from earth, is now considered a possible source of the WOW signal. As such, researchers from SETI feel this would be a prime target in the search for technosignatures. These are markers of technology which could represent intelligent life forms inhabiting exoplanets.

Techno-signatures typically consist of chemicals within the atmospheres of these exoplanets, such as oxygen or perhaps phosphine gas. The latter was recently discovered within the clouds above the surface of Venus[36] by scientists in the year 2020. Other signatures include light, particles, and gamma-ray emissions. The terms biosignatures and technosignatures are used interchangeably; however, the former does not necessarily indicate intelligent life. Conversely, the term technosignature does indicate the presence of intelligent life. Thus, the WOW signal, if proven to be genuine, would indicate the presence of a technologically advanced civilization which is capable of transmitting narrow-band radio signals.

More recently, SETI has instituted machine learning in the search for techno-signatures which was first reported in the journal Nature[37] in 2023. Here, researchers discovered eight possible techno-signatures using a machine-learning model. These eight signals have provided evidence of an origin between 30 to 90 light years from our solar system and seem to be propagating from five-star systems within that range. Although researchers are confident that these candidate signals have characteristics associated with ET technology, such as narrow-band frequency, non-local acceleration, and the appearance

---

[36] https://news.mit.edu/2020/life-venus-phosphine-0914
[37] Ma, P.X., Ng, C., Rizk, L. et al. A deep-learning search for techno-signatures from 820 nearby stars. Nat Astron 7, 492–502 (2023). https://doi.org/10.1038/s41550-022-01872-z

in strictly ON-source observations[38], unfortunately, the signals which were originally detected in 2016 have yet to repeat.

In the event that an ET radio signal is authenticated, the next step might be even more difficult with regard to deciphering the contents of the communication embedded within the signal. Morrison et al. (1977) proposed a five-step procedure to decipher and/or decode signals from extraterrestrial civilizations. The first step is termed the preliminary analysis of the signal which is directed toward the goal of establishing an alphabetical blueprint of the signal. The second step deals with how to determine the organization of the language which can be pictorial, linguistic, binary, algorithmic, or perhaps even mathematical.

In chapter five, we explored the pictorial images displayed in crop formations; hence, this would be an example of pictorial language if we are to assume that crop formations have ET origins. The next step would involve the determination of grammatical structure, and rules of the language construction. Perhaps we may now begin to see a relationship to some earthbound languages.

Next would be an investigation and determination of the semantics represented by the language. This may prove to be the most difficult step since it is quite

38 Terrestrial radio interference typically occurs in both ON and OFF observations because the source of the transmission is close to the receiver.

possible that ET civilizations convey meaning to language of ideas which are totally beyond our scope of cognition or for which we are not able to achieve conscious awareness. The last step would involve the translation of the ET language into known earthbound languages. Again, however, this could prove challenging considering the possible obstacles outlined in the previous step. Other limitations when it involves simply decoding a detected ET signal include the fact that we will most likely have no physical representation of the beings that are transmitting these signals. Thus, we are not able to ascribe emotional context to the messages being sent, assuming these beings have comparable emotions to human beings. It is possible they are void of any recognizable human emotional states, more akin to AI/robotic rather than biological. Another consideration is communication via mathematical concepts, which is highly likely since math is presumed to be a universal language. However, aside from conveying information about trajectory, coordinates, and raw data-driven information, this type of communication would do little to convey attributes such as emotion, instinct, or states of consciousness.

The search for life in our universe may even be closer than we think and may have been already discovered in 1976 with the "Labeled Release" experiment, designed by NASA researchers Gilbert Levin and Dr. Patricia A. Straat, as part of the Mars

Viking Lander mission[39]. These were early robotic landers that investigated the surface of Mars and analyzed soil samples to look for any signs of biology. Although some of the results were consistent with the metabolic processes one would expect to see in biology, the scientific community remained skeptical. Interestingly, although the functionality of the experiment was robust, on the many subsequent lander missions to Mars, a replication of this experiment has not been attempted. The experiment was tested extensively to rule out the possibility of false positive results occurring prior to the mission. These tests occurred in the most harsh environments on earth, from Death Valley to the Antarctic region, in both fertile and sterile soil samples, and each time, the results matched accordingly. Hence, no signs of tagged radioactive carbon-14 byproducts were produced in the sterile soils, whereas a positive byproduct result was detected in the fertile soils.

The results of these earth-bound tests held up accordingly when the test was performed in Martian soil. Although separated by a distance of 4,000 miles, both of the Viking landers produced positive results along with appropriate results from control samples of soils which were obtained from underneath a rock, which produced a positive result. Furthermore, another control sample was sterilized by exposure to a 160 degrees celsius heating procedure. Accordingly, this produced a negative result. Lastly, the apparent

[39] https://phys.org/news/2016-10-year-old-viking-life-mars.html

metabolism produced by this unknown Martian organism proved to be quite fragile, according to Levin and Straat, since a significant reduction of metabolic activity resulted from heating the soil sample to only 50 degrees celsius, and, conversely, lowering the soil sample to 10 degrees celsius. This pattern of metabolic variability with exposure to temperature extremes was indeed some of the most robust evidence of biological life because it demonstrates the Goldilocks zone effect on a microscopic scale.

Although the labeled release experiments do provide tangible evidence of microscopic life on Mars, there are some researchers, such as Percival Lowell, Elias Colbert, Dr. John Brandenburg, and Richard Hoagland, who have proposed the possibility of ancient technological civilizations on Mars. Some researchers have speculated that, somehow, this ancient civilization seeded earth and gave way to modern humans. There are other researchers who have speculated that modern humans were brought to earth from another distant world, such as Alpha Centauri B, which is a presumed habitable planet orbiting within our closest neighboring star system, approximately 4.25 light years from earth (Silver, 2017). This is a triple star system, which is about 6.5 billion years old. In comparison, our sun is 4.5 billion years old. In the Centauri system, strong solar winds may not produce habitability since any orbiting exoplanet would be close to these stars which are in the red dwarf category.

Red Dwarf stars are notorious for blasting off robust solar flares, which are typically not associated with providing a hospitable environment for life to evolve. Wherever the origin of human beings turns out to be, one can be certain that the theory of human evolution occurring entirely on this planet is being called into question with greater frequency and with stronger arguments as we have discovered more and more gaps in this supposed timeline and problems with evolution itself. For instance, there is a big problem when one examines the Cambrian period, which occurred 530 million years ago. This was a period of time when there was an explosion of new body morphologies which suddenly appeared on earth with a paucity of Precambrian (precursors) to explain this sudden appearance of novel life forms (Meyer et al., 2003).

It turns out that below the sediment rock layers of the Cambrian period, a paucity of fossils is apparent, and the few that are found do not show a clear and precise pathway leading to the complex morphologies found in the Cambrian period, such as Starfish, vertebrate fish, brachiopods, and arthropods. See, evolutionary theory dictates a slow and steady process that progresses with small steps and small improvements with the appearance of each subsequent life form. In contrast, the current trajectory from Precambrian to Cambrian proceeds from very simple to complex life forms without any intervening steps to account for the evolution of this complexity. In fact, it is like someone basically transported a plethora of complex animals to earth and dropped them off here, much like one stock the shelves at a grocery store. Let us speculate on this for a moment. Perhaps intelligent beings have prepared our habitable world in a manner such that plant life and animal life were purposely brought here so that a fertile environment exists for the benefit of the eventual humanoid civilization which will inhabit this planet. Thus, it is our responsibility to be kind to ALL of our fellow earthlings!

References:

Harold P. Klein, The Viking biological experiments on Mars, Icarus, Volume 34, Issue 3, 1978, Pages 666-674, ISSN 0019-1035, https://doi.org/10.1016/0019-1035(78)90053-2.

https://www.sciencedirect.com/science/article/pii/0019103578900532

Morrison, Dr. Phillip, Dr. John Billingsham, and Dr. John Wolfe, eds. The Search for Extraterrestrial Intelligence. NASA SP-419, 1977, 276 pages. The discussion about interstellar travel is absurd.

https://phys.org/news/2016-10-year-old-viking-life-mars.html

Meyer, S.C., M.R. Ross, and P.A. Nelson, P. Chien. 2003. The Cambrian Explosion: Biology's Big Bang. In Darwinism, Design, and Public Education, edited by S.C. Meyer and J.A. Campbell. Michigan State Press, 336.

Silver, E. (2017). *Humans are not from Earth.* ideas4writers.

Chapter 7
The Operations Manual

A fresh perspective

One of the most compelling aspects of the UFO subject is the existence of a leaked operations manual which serves as a step-by-step guide detailing how to manage all aspects of a militarized UFO crash retrieval operation. This document was originally leaked to UFO researcher Don Berliner in March 1994 by an unnamed source after he had attended an experimental aircraft exposition in the city of Osh Kosh, Wisconsin. The operations manual was not provided as a hard copy, but it was baked into a roll of 35 mm film and sent to Mr. Berliner in a manilla envelope with no return address. Interestingly, the individual frames of the film included the shadow of a human hand turning each page of the manual, recorded onto each individual frame of the film. Later on, the manual was transcribed by other researchers, most notably Stanton Friedman, Dr. Robert Wood (Physicist), and Ryan Wood. This manual has been written about in several books throughout the history of ufology.

The recent congressional hearings and the revelations provided by military and government officials regarding the existence of retrieved alien craft and biological materials bodes well for the credibility and legitimacy of the operations manual, which has been the subject of heated debate among researchers and skeptics for the past 30 years.

One simple question is salient upon review of the operations manual: Why would someone go through the trouble of creating a document such as this? Indeed, this manual is densely packed with information regarding the nuts and bolts of how to manage the successful retrieval of a downed alien craft and its occupants[40]. The manual includes everything from how to handle the material in terms of packaging, cataloging, and processing and even includes information about how to maintain secrecy surrounding each individual incident, including witness management and management of media interest.

Furthermore, the intricate details of the document have been studied by researchers Richard Dolan, Stanton Friedman, Ryan Wood, and Dr. Robert Wood in terms of everything from font size and style and the most likely region of storage and publication. Compelling evidence has been researched and reported by Ryan Wood to be consistent with Kirtland Air Force Base in New Mexico as the origin and storage facility of the manual. Identification of the printing press used to print the manual and verification of the initials of military officials who

---

[40] For a complete reproduction of the operations manual, see pp. 167-184 of TOP SECRET/MAJIC.
Operation Majestic-12 and the United States Government's UFO Cover-up. By Stanton T. Friedman, MSc. Copyright 1996, 1997, & 2005.

regularly reviewed the document have been supported by the meticulous archival research of Stanton Friedman, Dr. Robert Wood, and Ryan Wood. In short, when one makes a list of consistencies versus inconsistencies when trying to determine the legitimacy of this document, the former far outweighs the latter. In fact, most skeptics have not provided any reliable evidence showing this document to be a hoax other than a proclamation of such. Furthermore, skeptics seemingly overlook or fail to review the entire document in detail before proclaiming it as a hoax, which is a common response to any historical evidence which is brought forth in this field. Conversely, researchers in the field of Ufology have proven to be meticulous and guided by scientific principles when evaluating evidence in this field. Conclusions are not given as fact but as a likelihood based upon the preponderance of available evidence. The manual can be seen reproduced on numerous websites and blog posts with a quick Google search for those interested. Let us review some of the highlights from the manual which I will present within the context of current UFO/UAP revelations.

The date on the manual is April 7th, 1954. Section one of the first chapter is split into five parts which are numbered as subheadings as they pertain to MJ-12 and the operation manual.

The manual describes the properties of craft, which were limited to craft recovered during the years 1947-1953. Thus, many of the descriptions of the craft are quite bland, limited to properties such as color, texture, consistency, etc., presumably because, at that period of time, our scientific capabilities were not advanced enough to determine any intricate details about these craft and/or the occupants. The subheadings included in the first chapter of the operations manual include the following: *Scope, General, Security Classification, History, and Current Situation.* The highlights of this first section are to provide a historical reference for the MJ-12 group[41] and rationale for the fact that the information described herein is of the highest order of top secret for the reasons of public protection and safety (presumably to prevent the breakdown of the public trust and peace of mind). This has always been held as a top priority based upon the premise that knowledge of the existence of extraterrestrial (ET) life would cause widespread civil unrest and religious upheaval.

This belief has been held firm ever since the Brookings Institute published its findings in 1960 and basically determined that official disclosure would cause significant civil unrest and the possible destabilization of society. Additionally, this section of the manual states that the top-secret nature of their work is necessary to prevent the knowledge obtained from the recovery of ET hardware and biologics from falling into the hands of foreign adversaries. Hence, knowledge of advanced technologies by our enemies

---

[41] MJ-12 was established in September of 1947 and reported directly to the president of the United States, Harry S. Truman.

could offer an unfair advantage on the world stage in terms of military prowess. The fact that the original jurisdiction of MJ-12 was under the umbrella of the executive branch is interesting since now it seems that most modern presidents are not privy to detailed information regarding recovered tech and biologics. In fact, it appears that some time during the subsequent 77 years, the president was relieved of his direct knowledge of the history and current status of these recoveries. Some researchers have even suggested that the president is not even routinely read-in to these programs which require the highest level of security clearance.

Interestingly, the first section notes that ET entities are visiting earth with benign intent and not interacting with humanity. In the years since the operations manual was leaked, it should be noted that more than a handful of abduction reports have occurred, along with stories of alien implants and cattle mutilation cases, which have also garnered extensive attention. This would be an indication that the authors of the manual were unaware of these other related phenomena at the time.

The second chapter of the operations manual largely details the characteristics of the recovered craft, details of the inner workings of the technology including metallurgical analysis and engineering aspects, as well as descriptions of the entities recovered from the craft. The shapes described include disc-shaped, cigar-shaped, oval or circular, and triangular. Interestingly, the manual does not provide a description of oddly shaped craft as that which has been described in the past 10-15 years of cubes with spheres. Perhaps this most recent variation represents evolving technology or the appearance of a different species of ETs that are now surveilling our planet. The manual documents that recovered craft have been anywhere from 30 feet up to 300 feet in greatest lengthwise dimension with the exception of cigar-shaped craft, which are documented to be 2000 feet long and 95 feet thick according to data which was ascertained by radar data at that time. Based upon this admission of data regarding dimensions of the cigar-shaped vehicle being ascertained by radar data solely, one might conclude that crash retrievals of cigar-shaped vehicles did not occur during that span of time (i.e. 1947-1953).

Furthermore, a review of the available literature does not result in any credible evidence of crash retrievals of cigar-shaped craft. Furthermore, the operations manual states that detection of the cigar-shaped craft is rare and limited in scope to radar data, which indicates that operation of these craft is limited to high within the atmosphere and with relative speeds greater than 7000 miles per hour. As such, one could determine that crashes due to low atmospheric conditions, catastrophic contact with low atmospheric military hardware, airborne obstacles, or ground-based ordinance are highly unlikely, thus accounting for the paucity of evidence that these craft crashes with any regular occurrence.

The material and debris recovered during that date range (i.e., 1947-1953) is described as the appearance of aluminum-magnesium alloy with a high tensile threshold. In the manner in which the debris was described in the manual, it would appear that at least some of this wreckage was the result of military ordinance causing an impact and subsequent explosion, bringing the craft down. Still other craft have been recovered largely intact and have apparently either landed either voluntarily or forcibly through mechanical interference or malfunction. Critics and skeptics have argued that the suggestion of technologically advanced alien craft traversing tens of light years to visit earth to then just crash is laughable. However, they are not taking into consideration that humans are a technologically advanced species and a small percentage of our aircraft crash on a regular basis. They are, hence, overlooking the fact that no entity is totally, 100% immune to accidents, no matter how advanced they are.

The critics and skeptics make the very big mistake of failing to look at things from a *practical* perspective. They assume that non-human entities possess infallibility and this is quite obviously not the case. Nothing in the universe is immune to injury and death. Stars are born and die; planets are born and destroyed by asteroids; entire galaxies are ripped apart and collide with each other. Furthermore, all of the non-human and human species on earth are susceptible to accident, injury, and death whether it be from natural causes, predators, accidents, or disease. Why would advanced alien civilizations be any different?

The second chapter also describes some details about two types of beings that have been recovered from downed craft and provides basic details about their body habitus. In the operations manual, they are referred to as Extraterrestrial Biological Entities (EBEs). The first type is described as bi-pedal, approximately 80 to 100 lb, and between 5 to 5 '4 in length with similar proportions to the human body except for the cranium, which is larger and with greater roundedness, particularly along the parietal and occipital regions. The hands and feet are without webbing and contain only 4 digits, and the skin is yellowish, thick, and pebbled in appearance. They have a paucity of body hair and the ears are smaller than that of humans, while the mouth is wider and the nose is thin and longer. The eyes are the shape of almonds with black irises and large pupils, set wider apart than that of humans.

The second type of EBE is described as bipedal, approximately 25 to 50 lb, with a relative length of 3 '5 to 4' 2. The skin is pale blue, they have no visible ears, the nose and mouth are slit-like and basically flush with the surface of the face, and their craniums are much larger and elongated compared to both humans and EBE type-1. The bodies are without hair and the fingers number three plus a thumb. There is a question about whether or not these are mammalian beings, mainly due to the fact that the arms are very long with respect to the legs. There are four toes on the feet which are connected to each other with a webbing-type membrane.

There has been speculation among various researchers about whether or not some of these beings are constructed with non-biological components such as artificial intelligence. However, one can see how the distinction between biology versus non-biological components might be blurred, especially considering the fact that humans are now at a point where we are able to install components into the human body that basically replicate the function of our biological parts. I think the more salient question deals with whether or not these beings have a built-in consciousness and whether or not that consciousness contains some form of ethics, morals, and caring for the well-being of other beings.

The next section of the manual deals with the nuts and bolts of security operations when it involves securing a UFO crash site. It is important to note that the operations manual uses the acronym UFOB when referring to unidentified flying objects instead of the modern nomenclature of UAP (Unidentified anomalous phenomena). This chapter is split into seven sections, which include how to manage the media, security of the immediate area, maintaining an inventory of recovered ET artifacts, proper packing and packaging of materials, removal/transport of materials (including biological materials), and lastly, cleaning the area.

During recent senate testimony by the head of the All-Domain Anomaly Resolution office (AARO), Dr. Jon Kosloski and his predecessor, Dr. Sean Kirkpatrick, were called upon to deliver a status report on the 1,652 cases they are charged with investigating. Thus far, the position of both of the heads of these agencies has been less than revealing with regard to methods they are using to investigate with the general public, civilian investigators, and members of Congress. Additionally, they seem to have taken a page right out of the operations manual "playbook" for how to deal with the public and attention from the media but now have expanded this to the congressional officials who are asking questions. This is particularly true with regard to providing statements that localize the etiology of UAPs to mundane objects such as downed satellites, balloons, meteors, military aircraft, and weather phenomena. Furthermore, no attempts by AARO have been made to speak to or seek guidance from experienced civilian investigators or any of the reputable civilian agencies such as the Mutual UFO Network (MUFON), Center for the Study of Extraterrestrial Intelligence (CSETI), and the National UFO Reporting Center (NUFORC). Also, efforts to meet and coordinate with military witnesses have been a non-starter. As such, it is apparent that the same reports of compartmentalization plague

AARO as was reported by Robert Lazar during many of his interviews over the past 35 years.

The operations manual documents many of the techniques we have all become familiar with which include discrediting witnesses, official denials, and deceptive statements. The most damaging of these techniques is the common practice of blaming the witnesses for hoaxing events or suffering from hallucinations or hysteria. This last technique has prevented many witnesses from coming forward as is the case of pilots who are concerned about losing their jobs and livelihood.

As far as the internal biology of the non-human entities is concerned, this is left to much speculation. There have been reports of apparent autopsies which have been performed on the recovered bodies. There are reports from anonymous sources that autopsies of non-human beings have uncovered the existence of one solitary organ which performs the activities of circulation, breathing, and digestion. Other reports have noted a diminished capacity of these beings to rid themselves of waste products. As such, it has been offered that perhaps this is the reason for some abductions and cattle mutilations, i.e., to harvest biological products such as tissue samples or bodily fluids, which are then used to grow usable biological materials in-vitro, which can eventually be used by alien beings to repair deficiencies or enhance their inefficient anatomical and/or physiological functions.

The authenticity of the operations manual has been questioned by both skeptics and some UFO researchers throughout the years since it was first revealed. In this author/researcher's opinion, one of the most obvious aspects of the operations manual, which supports its legitimacy, is the emphasis on mundane details. In this regard, the manual is certainly not written for the reader's pleasure. In fact, some sections are about as boring as watching paint dry. As such, it seems to have certainly been written for militarized teams whose sole job is to secure and clean the immediate site of a crashed UFO/UAP. This mundane theme is especially apparent in the sections concerning the cleaning of the crash area, packaging of material, unpacking of material, and managing the debris check-in and inventory process. These sections of the manual are separated into the handling of non-biological and biological material. The amount of detail in these sections is clearly intended to achieve the goal of debris and data preservation. Some of the instructions provided include using moisture and vapor-proof barriers within the shipping containers along with packing materials that restrict the shifting of continents, including neutral cellulose wadding corrugated fiberboard boxes sealed with gummed Kraft tape. With regard to the unpacking of extraterrestrial specimens, technology, and debris, the operations manual outlines a detailed 13-step process. Again, we must ask ourselves why anyone would choose to hoax a document that includes such mundane details. I have summarized these steps herein:

1) Before unpacking shipping containers, verify you are in a top security area without access to unauthorized individuals.
2) Metal straps which are holding shipping containers closed shall be cut with a wire cutting tool or twisted by hand until the straps break.
3) Any screws securing the shipping container should be removed with a screwdriver.
4) Cut any tape which is sealing the liner being careful not to damage waterproof material.
5) Carefully lift and remove the packaged materials from the wooden container.
6) Carefully cut the tape that secures the flaps which are atop the lateral cartons; do not damage the cartons.
7) Remove the inner carton by cutting the barrier on top of the thermally-sealed seam.
8) Carefully extract the envelope which sits atop the innermost carton.
9) After opening the innermost carton, carefully remove the fiberboard, desiccant, and humidity monitoring device.

10) Remove the thermally-sealed packaging material which contains specimens and carefully arrange them for inspection.

11) Save all of the packaging material for future use within the shipping container.

12) Thoroughly take inventory of all specimens by comparing them against the shipping documents inside of the envelope, which was removed in step eight.

13) Personnel with proper security clearance are responsible for the covered transport of specimens to authorized laboratory departments. Security guards will also be present during this transport process.

The recurring theme in the operations manual is the fact that all transactions are done under high security and are performed extremely carefully. This practice evolves into a highly compartmentalized process whereby, eventually, all specimens and materials are assigned to isolated, small groups of 2-3 individuals. These teams have limited or no contact with other teams because this increases the chance that information will leak out into the public realm. The common practice of compartmentalization within government or military operations enables these authorities to easily track which team and/or individual the leaked information originated from.

The final sections of the operations manual deal exclusively with how to coordinate the handling of extraterrestrial organisms (both living and nonliving) and some basic rules to follow for personnel who are in charge of investigating UFO sightings. The manual makes it clear that contact with extraterrestrials is to be denied under any and all circumstances to the general public and to anyone without proper security clearance. The manual breaks down encounters with extraterrestrial biological entities (EBEs) into two sections: those that are initiated solely by EBEs and those that occur due to grounded alien craft. In the former circumstance, the manual does acknowledge the possibility that contact initiated by EBEs could occur in a non-controlled setting. In the event this occurs, any and all effort should be made by the military to secure the area. The operations manual supports the use of intimidation and discrediting tactics as acceptable methods of ensuring witness cooperation. The fact that witnesses to UFO sightings have been largely ridiculed for 80-plus years lends support to the legitimacy of the operations manual because it clearly documents the acceptance of those practices. By doing so, it clearly set a precedent which has proven robust and has infiltrated both legacy science and popular media.

Some interesting details in these last sections include the fact that reference is made to administering medical care to injured EBEs. This care is mostly directed at stabilizing injuries, which includes controlling bleeding, bandaging wounds, and splinting damaged limbs. The manual makes a point to mention that no medications are to be given to

EBEs due to the fact that the effect on nonhuman biology is unpredictable. Again, the importance of maintaining security is stressed in this section, which notes that, above all else, maintaining security is paramount, even if the life of an EBE is in peril. The manual references specialized personnel on several occasions whose responsibility is to provide care, security and custody to the EBEs. Other details include the importance of wearing protective attire, including surgical gloves, and limiting the use of bare hands to handle biological specimens. The importance of decontamination after the handling of specimens is stressed in the manual in light of the fact that EBEs may carry pathogens which are toxic to human beings. Lastly, the manual references the importance of the preservation of biological specimens to prevent decomposition. This includes placing specimens in specialized containers, which are then time stamped, dated and then either submerged in ice or refrigerated for preservation purposes.

As stated previously, the authenticity of the operations manual is inherent because of its deep dive into mundane details. Most of the content of the manual is not "exciting" per se. It is very much a working manual that is intended to be used by specialized teams ordered to carry out specific tasks. The circumstances surrounding the appearance of the manual in an unmarked envelope, the fact that it was baked into a roll of 35 mm film, and the appearance of a solitary hand within those frames, which is holding each page of the manual such that it can be recorded onto the film, all add to the mysterious nature of this manual. None of these features prove its authenticity. However, when taken all together, within the context of recent whistleblower revelations and 2023-2024 congressional hearings, the authenticity of the operations manual appears to be genuine, in this author's opinion.

Reference:

TOP SECRET/MAJIC. Operation Majestic-12 and the United States Government's UFO Cover-up. By Stanton T. Friedman, MSc. Copyright 1996, 1997, & 2005.

Chapter 8
Facts, Figures, And Researchers

In the following pages, we explore the brick-and-mortar facilities and personnel which are part of the traditional lore when it involves the secrecy of UFOs. Some of these include Area 51, Wright-Patterson Air Force Base, Frank Scully, Stanton Friedman, Robert Lazar, and John Lear. We will explore these and some you may not have heard of. When reading this information, try to keep an open mind and ask yourself the following questions: What is the perceived credibility of the person? What does this person stand to gain by being deceptive or hoaxing any of the information they have provided? Does the person's story fit into the larger picture of the UFO phenomenon in such a way that it adds or compliments information that is already known to be credible, as determined by corroboration by seasoned researchers in the field? Moreover, keep in mind when reading that witnesses involved in the subject area of UFOs have taken quite a risk because of the stigma that is attached to the subject.

As we touched upon in Chapter 7, the operations manual lays out a blueprint for dealing with witnesses and how to basically discredit them. This involves accusing them of deception, being in a mind-altering state such as suffering from hallucinations, or even worse. This is indeed an effective method of keeping most people quiet. However, as we'll see, a few of these individuals are still willing to go out on a limb and tell their story, in some cases risking everything for the sake of promoting a small microcosm of disclosure.

One of the goals of this book is to impart certain basic knowledge and temporal data with regard to the UFO phenomenon to the reader and prompt them to ask themselves, what is the probability that all of this has occurred by chance or that a concerted effort to hoax this phenomenon has been undertaken by the thousands of witnesses and researchers entrusted with its historical integrity. Three paramount events which triggered the modern era of ufology should be noted; all of these occurred during the summer of 1947. The first event occurred on June 21st, 1947, in Tacoma, Washington state and involved resident Harold Dahl, his son, and the family dog, who were on a boat traveling across Puget Sound to Maury Island to collect logs. Upon reaching the shore of the island, they observed six flying discs hovering above the shoreline. One of the discs appeared to be having some type of mechanical difficulties as it appeared to be in an uncontrolled yaw as it slowly moved towards the boat.

Once directly over the boat, the disc began to drip some type of molten hot liquid onto the boat which hit the family dog and Harold's son. The dog later died as a result of the injuries and his son needed treatment at a local hospital. Harold promptly reported the incident to local authorities and an investigation ensued. The next morning, Harold received a knock at his front door by an official dressed in all-black attire (i.e., the origin of the "Men in Black" phenomenon). This official warned Mr. Dahl to avoid discussing the incident any further. Although the official did not give any direct threats, Mr. Dahl felt very intimidated by the man and as a result, he recanted his statements a few days later. Skeptics claim Mr. Dahl made up the entire story; however, in light of the vast reports throughout history of UFO witnesses being harassed by anonymous authorities, the veracity of Mr. Dahl's original account is a sound bet. For 50 years after the incident, the FBI file on the Maury Island incident was sealed. Upon its opening, it was discovered that Mr. Dahl only recanted his original story to avoid negative attention and consequences. However, in private and off the record, he stood by his original account of the incident. Interestingly, just 3 days after the Maury Island incident, pilot Kenneth Arnold witnessed nine discs flying near Mount Rainier, also in Washington State. The commencement of "The Summer of the Saucers" in 1947 concluded with the August 8, 1947, Roswell, New Mexico flying saucer crash which has unarguably become the most well-known incident in history.

Frank Scully (1892-1964) was an early author with access to credible information regarding the UFO phenomenon. He was one of the first authors to popularize the notion that flying saucers utilized some sort of magnetic propulsion system to propel themselves in the earth's atmosphere. Moreover, he was the first investigator to report on the alleged Aztec, New Mexico flying saucer crash of 1948. At the time of his publications, the skeptics highly criticized him because some of his sources were accused of being involved in some illegal activities. However, when considering Mr. Scully, one is cautioned not to disregard all of his claims because of the shortcomings of one source. This certainly does not mean all of his claims are untrue. He was the first to popularize the crash retrievals of flying saucers in both New Mexico and Arizona, which in today's context are certainly held in a new and credible light, especially with the recent testimony of whistleblowers who have testified before Congress in 2023 and 2024. Interestingly, through Scully's sources, he was one of the first to promote the notion that lines of force similar to electrical current existing outside of a wire are what is responsible for this propelling force of the crafts. In his 1950 book, "Behind the Flying Saucers," Scully states that his source disclosed that the motion of flying saucers is accomplished through "reverse electrical energy" such that the saucer is always trying to flee away from disturbed magnetic foci, similar to the way a person tries to run away from someone trying to poke them with a needle.

Surprisingly, recent reports (some of which have been given under-oath to Congress) by military personnel, scientists, aerospace professionals, and astrophysicists have asserted that these anomalous crafts indeed travel via a propulsion system which does not utilize combustion of fuel, but instead (most-likely) uses reverse-gravity technology. A particular idea that Scully alludes to is the notion that since flying saucers utilize magnetic lines of force for propulsion, this significantly increases the velocity by which these vehicles can travel both within the atmosphere of a planet and assumably outside of the atmosphere of a planet. In the latter example, the paucity of friction creating resistance is null, thus negating any limits upon velocity.

As such, Scully provides the example that a hypothetical spacecraft using magnetic force for travel within the interstellar medium could theoretically achieve a trip from Venus to Earth (or vice-versa) at the speed of light (186,000 miles per second) to achieve this trip within 42 minutes. Scully also talked about the apparent intolerance of alien beings to the earth's atmosphere by speculating that this was the etiology of their demise in the reported saucer crashes and recoveries during the 1940s and 1950s (i.e., Aztec, NM, Kingman, AZ, Roswell, NM). As such, Scully promoted the theory of decompression as being responsible for the deaths of these beings upon entering the earth's atmosphere, similar to that which is experienced by scuba divers who come to the surface too quickly and suffer "the bends," which can result in death. Physiologically, what happens is that gas bubbles form within the bloodstream and can travel to vital organs such as the heart and brain and block small blood vessels through a mechanism that promotes the production of micro-emboli. This can cause respiratory failure, stroke, and myocardial infarction with subsequent cardiovascular failure. Assuming that the physiology of non-human beings is similar to ours, it is not difficult to speculate how they too might be susceptible to a similar pathological process. As we can see, Frank Scully's writings paved the way for three major ideas: UFO crashes, mode of propulsion, and the susceptibility of non-human entities to both injury and death. Although Scully's ideas are theoretical and are not meant to be taken with absolute certainty, they have accomplished something very important: a thought "zeitgeist" that exists to this day and has been expanded upon and extrapolated by credible scientists within this evolving field.

The first person connected to the government to investigate UFOs was Captain Edward Ruppelt. He is credited with the birth of the term Unidentified Flying Object and was in charge of the government's investigation of UFOs from 1951 through 1953, called Project Blue Book. Mr. Ruppelt served initially as the leader of Project Grudge, which was later named Project Blue Book when a more serious tone regarding the objective investigation of UFOs was put forth.

Unfortunately, the former investigation (Project Grudge) had only intended to debunk the UFO phenomenon. The most salient aspect of Ruppelt's work was the weight he placed upon radar-visual sightings of UFOs. Ruppelt stated that these were indeed the most convincing data points because you have positive ground observations, positive radar targets, and positive identification by pilots who are called to scramble these UFOs, oftentimes with subsequent pilot-initiated radar lock-on. As is typical, however, the UFOs typically outdistance the military craft within the blink of an eye, a maneuver which, even until present-day, is not within the capabilities of any known earthbound aircraft. Like any credible investigator, Captain Ruppelt approached the UFO subject from a position that assumed the "null" hypothesis[42] to be, in fact, true. However, after years of investigating UFO reports, he accepted the fact that a small number of cases had no apparent explanation.

Stanton T. Friedman (1934-2019) was an American-Canadian Physicist who served as a major authority figure in the field of Ufology. He was a vocal advocate of the scientific study of the field, and as such, he tirelessly traveled the world to lecture to small groups, large groups, and individual people about his research for over 50 years. Forever considered an authority in the field, Mr. Friedman held strong beliefs that alien craft and beings have been visiting the earth regularly throughout the 20th century and what sparked this influx of surveillance was the fact that humanity had begun nuclear weapons programs and commenced regular detonation of nuclear weapons both the Southwest,

---

[42] The "null" hypothesis states that any evidence in a data set which seems to support the hypothesis is merely a result of chance, sampling error, or experimental error.

United States and in the former Soviet Union.[43] Mr. Friedman was the original civilian investigator of the Roswell incident and also wrote extensively about the Betty and Barney Hill abduction case and MJ-12. Throughout his tenure, and on the basis of his capacious archival research, Mr. Friedman held strong opinions in the direction of an extensive UFO cover-up, which he believes has been propagated for the majority of the 20th and 21st centuries and maintained by a small number of government insiders with ultra top-secret clearance. He took issue with other researchers on this issue, specifically J. Allen Hynek, who contended that the government really doesn't know what UFOs are.

Conversely, Mr. Friedman always insisted they do know what they are indeed and referred to the cover-up as the "Cosmic Watergate." By using this term, Friedman drew similarities to the political conspiracy that eventually brought down President Nixon and forced him to resign from the office of the Presidency in 1974 because of the ongoing impeachment process which had commenced in the House of Representatives. Adding to the credibility of Stanton Friedman was the fact that he was not a researcher who would believe every case which was presented to him. He was a meticulous factfinder which involved eye witness interviews, archival research, and scientific and data analysis. His research relied

---

[43] These nuclear tests began what is referred to as the "Cold War" between the United States & Russia
(formerly the Soviet Union)

entirely on the embracement of the scientific method and the piecing together of data with relevant archival and historical information. He lectured at colleges, universities, and professional organizations all over the world. His credibility was also bolstered by the fact that in his early career, he worked as a nuclear physicist with a top-secret clearance for organizations like General Electric and McDonnell Douglas.

One of the major themes of Stanton Friedman's career was his persistent criticism of the flawed conclusion put forth by the 965-page Condon report (Scientific Study of Unidentified Flying Objects) completed in 1968 and published in 1969. The Condon report was the result of a 500K funding provided to the University of Colorado by the Air Force. The director of the study was Dr. Edward Condon and its goal was to study the scientific veracity of UFOs. Although much enthusiasm was apparent at the beginning of this study, it soon became apparent through leaks of committee members that the study's ulterior motive was to throw cold water onto the subject of UFOs by purposely spinning the data such that Dr. Condon's conclusion was basically that there was "nothing to see" and that UFOs are not worth investigating. His conclusions ran in stark contrast to the actual data buried in the 900-page report which demonstrated a whopping 30% rate of unknowns of the 550 cases available for study within the Air Force database. The Condon report basically cherry-picked explainable cases and ignored cases which defied explanation, leading to the foregone conclusion that UFOs do not exist.

At the time, this conclusion was quickly endorsed by the National Academy of Sciences. As such, it was erroneously cemented into the historical record as the "final word" on UFOs, and sadly, this stigma held for almost 50 years among a significant number of individuals and most governmental bodies and fueled by the popular media who constantly poked fun at UFO reports and snickered whenever they reported on reported UFO sightings and tales of alien contacts and/or abduction.

Furthermore, any time a researcher was interviewed, he or she was always challenged by a skeptic or a debunker who typically did not spend any time or effort actually researching the cases or subjects. They simply threw cold water onto any claims of UFO sightings or contacts by proclamation of a hoax, a hallucination, or confusion with natural processes. Some of the absurd causes of UFO sightings referenced by the skeptics included weather balloons, flares, swamp gas, hole punch clouds, and birds. The swamp gas explanation took on a life of its own as this seemed to have triggered a nerve with the general public who truly felt "gas-lighted" by this explanation by the government. The history behind this is a sighting which occurred in Dexter and Hillsdale, Michigan, in 1966 and was witnessed by many individuals in those communities (see Figure 1).

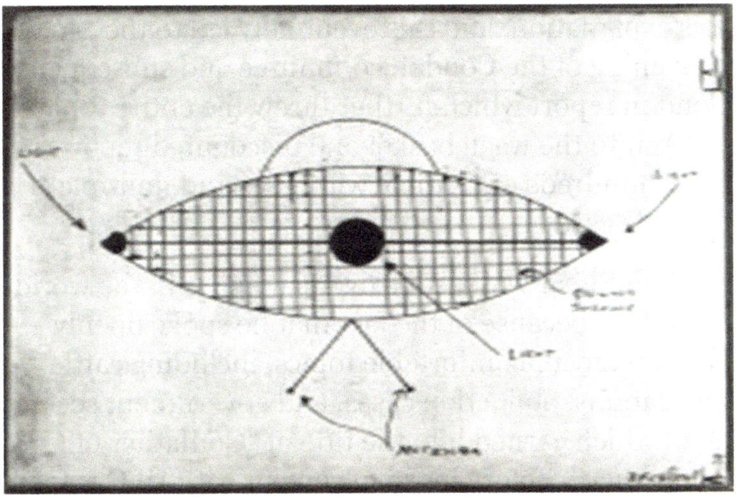

Figure-1. Credit to:
https://dexter.lib.mi.us/events/we-are-dexter-not-swamp-gas-the-true-story-of-the-1966-michigan-swamp-gas- ufo/

Residents of those communities, including college students, EMS workers, law enforcement officers, and other professional people, reported seeing brightly illuminated disc-like craft hovering in the evening skies. This was picked up by national news, and the congressman and future president at that time, Gerald Ford, called for an official investigation, which was headed by J. Allen Hynek of Project Blue-Book, who was a respected Physicist and Astronomer who held an academic post at Ohio State University. His official conclusion was that swamp gas was creating the illusion of these floating discs, subsequently misinterpreted by members of the public. Suffice it to say that the public felt disrespected and gaslighted by this explanation, and this eventually led to the convening of the Condon committee and subsequent Condon report which further threw the entire topic of UFOs into the wastebasket and discounted reports from hundreds of credible witnesses and multiple credible cases.

John Lear was a controversial figure in the world of ufology because of the fact that he spoke openly about many uncomfortable topics, including cattle mutilations, abduction cases, and government secrets, all of which earned him the title of "Godfather of conspiracy." Before his involvement with UFOs, John Lear had earned his place in history as a great aviator. He was a CIA pilot and was the holder of 17 world speed records. Furthermore, John was the only pilot in history to capture all of the different types of airman certificates awarded by the Federal Aviation Administration.

Moreover, John Lear came from a famous family in the Las Vegas area, as his father was the inventor of the Lear jet. His sister Tina provided a genuine summation of her brother John when she eulogized him on March 30th, 2022, with the following statement, "John was an incredibly complex man - to some a complete crackpot, to a great many others a genius." Considering the recent revelations from whistleblowers who have recently testified in front of Congress, one may reasonably argue that the genius of John Lear is the more valid argument. In a series of interviews in which John Lear appeared back in 1987 hosted by investigative reporter George Knapp, he summarized his internalized knowledge in a succinct and confident manner. John appeared to have inside information regarding alien bodies which are apparently stored in top-secret cryogenic facilities. These bodies represented multiple races of aliens, he affirmed. John also was one of the first researchers, along with Linda Moulton Howe, to openly state that some of these alien races have nefarious intentions towards humans and other earthlings, which presumably accounts for some of the abduction cases and the thousands of cattle mutilations which have occurred over the past century. Although extremely controversial at the time, many of the things that John Lear stated have indeed taken on a life of their own and have since found supporting evidence both in context and circumstance. Thus, John Lear's contributions to the field must be given the credence they deserve, as they have helped advance ufology as a legitimate scientific and existential discipline.

Growing up in the 1970s, Robert "Bob" Lazar was a guy who enjoyed tinkering around with car engines and race cars, taking things apart and putting them back together with "enhancements," and loved the freedom which came with the free exploration of scientific applications. He was never much for the "academic" aspect of science and technology. He was the guy who loved taking scientific theories and putting them to practical application. His interests evolved as he reached adulthood, and his projects of interest grew into the production of hydrogen gas from water by using electricity, which he then used to power his Chevy Corvette. Successfully refining this process, Mr. Lazar is able to produce enough usable hydrogen which he then stores in four compact tanks to propel his Corvette almost 400 miles, which he states is comparable to a full tank of gas. This process and end product are safe, effective, and economical, according to Mr. Lazar.

Another project involved powering his vehicle with a jet engine, which actually got him recognized in a local newspaper. Bob is one of those rare people who has the ability to think outside of the box and apply his knowledge in an entrepreneurial sense. He has founded several businesses, including photographic processing in the 1980s, and more recently founding a company that builds scientific equipment for various industries. During the 1980s, Bob's talents and abilities for thinking outside the box were noticed by some key individuals in the scientific community, including Edward Teller (1908-2003), who is known for his contributions to the development of the hydrogen bomb. These connections enabled Mr. Lazar to attain employment at some of the most prestigious research facilities in the United States, including Los Alamos and, later, the Nevada test site. During that period of time, the United States was still involved in the "Cold War," and as a result, these government facilities were involved in research and development projects to strengthen and secure the country's defenses. Some of the major projects included nuclear weapon development, advanced propulsion systems for aircraft, and Ronald Reagan's Strategic Defense Initiative[44] (SDI), "AKA," "Star Wars." The following excerpt from the United States Department of State Archives, which are open to the public (see footnote below (3)) summarized the Reagan administration SDI program:

> *The heart of the SDI program was a plan to develop a space-based missile defense program that could protect the country from a large-scale nuclear attack. The proposal involved many layers of technology that would enable the United States to identify and destroy automatically a large number of incoming ballistic missiles as they were launched, as they flew, and as they approached their targets. The idea was dependent on futuristic technology, including space-based laser systems that had not yet been developed, although the idea had been portrayed as real in science fiction. As a result, critics of the proposal nicknamed SDI "Star Wars" after the movie of the same name.*

A major goal of the Reagan administration, the SDI program relied heavily upon research development at various sites throughout the United States. Many of these facilities are located in the Southwest and large swaths of Nevada, Arizona, and New Mexico. This was a pivotal time in the country,

---

[44] https://2001-2009.state.gov/r/pa/ho/time/rd/104253.htm

and Mr. Lazar's talents seemed to fit the bill with regard to what the country needed in terms of contributions to advanced technology. Bob was eventually recruited to work at a remote installation on the Nevada test site in a project which was explained to him as "advanced propulsion." Bob had previously worked at Los Alamos as a physicist which allowed him to start his preliminary training and introduction to the program right away. As he discusses in his 2019 autobiography, "Dreamland," Lazar was soon to be amazed by what he witnessed at the facility.

As a lifelong skeptic of the UFO phenomenon, Mr. Lazar was not prepared for what he witnessed at the facility which included flying saucers of all shapes and sizes and work that involved trying to analyze and reverse-engineer an exotic propulsion system which did not involve the combustion of fuel, but instead involved the artificial production of gravity which when directed in reverse, allowed for these saucers to fly seamlessly and with extreme velocities. One of the most compelling aspects of this propulsion system, which Mr. Lazar alludes to, is the fact that it appeared to be 100% efficient and, as such, did not produce any wasted energy or heat. This aspect is quite different from human technology which is notoriously inefficient within this context and thus always produces wasted energy in the form of heat. Thus, the system Bob was working on conveniently violated the second law of thermodynamics, which states that no system can convert all of its energy into another system and that energy (heat) loss will inevitably occur. Thus, this process ultimately results in increasing entropy (Hewitt, 2015).

One of the most compelling pieces of evidence is a 40 minute video that Mr.Lazar recorded in the late 1980s in which he discusses this novel source of interstellar travel and the basic science which makes it possible. In the video, he discusses the bending of space-time and how this enables interstellar travel within efficient time. Mr. Lazar explains that gravity is actually a wave and explains there are two types of gravity, which are described as gravity-A and gravity-B. He explains that gravity-A is the force which holds together all of the protons and neutrons in an atom.

Furthermore, gravity-A is much stronger than gravity-B and as such, gravity-A is what needs to be accessed in order to distort space-time. Specifically, he states that in order to accomplish this, a stable isotope of a heavy element (element 115) is needed. The power source of extraterrestrial vehicles, Mr. Lazar explains, is a small reactor which utilizes element 115 and converts it to element 116, producing antimatter that is enclosed within a tube and directed in a matter that results in a 100% efficient thermoelectric generator which has extreme power, resulting in the production of antigravity. This antigravity is then directed through the emitters at the base of the vehicle which then allows access to and manipulation of the space-time medium. The controversy surrounding Bob's claims about the existence of element 115 has been criticized by legacy scientists due to its unstable nature. This element does not occur naturally on earth and must be created in a lab. This was accomplished in 2003 by scientists from Russia and the United States and was subsequently given the name Moscovium. The properties of Moscovium rendered it unuseful and unstable because of its high radioactivity, with a half-life of merely 20 milliseconds. Mr. Lazar has noted on several occasions that stable isotopic forms of this element are possible, and one such stable isotopic form was used in the propulsion system he worked on.

It is important to note that among UFO researchers, there exist two camps with opposite opinions about Mr. Lazar. His story has been debated for years and most likely will continue to be a controversial topic within the field. My opinion on Mr. Lazar is that I do find him credible. Despite the criticisms which have been levied, his story has been consistent for over 35 years, and many details of his story have gained support over the years as more and more information has been released by whistleblowers and through the recent congressional hearings. Some of these details include the existence of gravity waves, modes of disc flight (i.e., omicron, & delta configurations, roll maneuver), verification of test-site security equipment (i.e., hand scanners), and the discovery of element 115. For more information on the Bob Lazar case, I would encourage readers to view some of the interviews that he has done over the years with Art Bell, George Knapp, and, most recently, Joe

Rogan. Many of these can be found through a simple Google or YouTube search.

References:

Hewitt, L. A. (2015). *Conceptual physics*. Pearson education.

Lazar, B., Knapp, G., & Abrams, B. (2019b). *Dreamland: An autobiography*. Tantor Audio.

Ruppelt, E. (1956). *The Report on Unidentified Flying Objects*. Cosimo Classics.

Scully, F. (1950). *Behind the Flying Saucers*. Henry Holt and Company.

https://2001-2009.state.gov/r/pa/ho/time/rd/104253.htm

Chapter 9
Research, Nomenclature, & Status Report.

The approach taken in this book to literature and research review is historically dense. The reasons for this are multiple and include the fact that historical research is often less prone to manipulation by modern methods of photographic manipulation. Today, with the common use of smartphones, there is a wide availability of applications which can easily manipulate photos and even place objects into photos deliberately. This renders the process of investigation much more time-consuming and prone to error. This was not true of the photos taken during the 20th century which were captured on 35 mm film and later on instant photos in the 70s. These were not easily manipulated and would require physical manipulation of the environment. Furthermore, historical research is much less infiltrated by an organized system of debunkers and "information manipulators," many of whom are compensated to deliberately muddy the waters! To complicate matters, nowadays, we have the emergence of artificial intelligence which is further complicating the investigative process and the integrity or "genuineness" of recovered evidence. Back in the 1940s and 1950s, the system of secrecy was still in its infancy, and thus, its complicated organization was not completely solidified.

Skeptics argue the lack of scientific research with regard to UFOs and other anomalous subjects. There is a twofold problem with that argument, however. First of all, since the anomalous subjects of UFOs, crop circles, abductions, cattle mutilations, etc., are not amenable to the preferred experimental design, which is held to the highest regard in science (i.e., the double-blinded, placebo-controlled, quantitative experimental design), often the research that does exist is ignored, discounted, or rejected outright. Thus, the majority of research on these topics has been isolated to books and a small number of underfunded journals, such as *The Journal of UFO Studies*. If it were not for the dedication of a small number of researchers, these anomalous subjects would not have evolved into their current state.

Hall (1980) describes the concept of UFO imprinting which the author contends is the most amenable to objective analysis. This imprinting includes effects of UFOs upon the physical environment and includes inanimate objects such as roads, bridges, and other surfaces, organic systems such as livestock, humans, other animals, and botanicals, and effects upon mechanical systems such as power grids, and electronics. For example, researcher Michael Schratt describes the incident whereby the Kingman Arizona craft made blunt-force contact with a portion of the Colorado river dam when recovery teams were trying to move the disc across the river on a barge. This is obviously something which could be verified and thus would be an example of imprinting. Apparently, this resulted in some possible surface (cosmetic) damage to the dam.

Another landing case which occurred on August 19th, 1965, involved the landing of a disc at Cherry Creek, NY, near a farm, which was witnessed by multiple individuals. The Electromagnetic effects of the disc produced deleterious effects on the farm's milk cows which resulted in decreased milk production (Lore, 1969, as cited in Hall, 1980).

The renowned optical physicist and UFO researcher, Dr. Bruce Maccabee, Ph.D. conducted an analysis which was published in the *Journal of Scientific Exploration* (Vol. 13, No. 2, pp. 199-211, 1999. See link in footnotes)[45] in which he analyzed photographic evidence of a luminescent disc which was captured in a photo taken in flight by a Royal Canadian pilot at a reported altitude of 11 kilometers. The case report included testimony from the pilot that he believed the object to be stationary at the time the photo was taken. Dr. Maccabee's analysis of the data involved the utilization of a standard photographic equation corrected for atmospheric attenuation effects. Using various data points, including zenith angle, optical thickness, object radiance, film exposure level, and optic transmission, Dr. Maccabee was able to determine the shape of the object as a thin disc, its altitude, and estimate of power radiation greater than a gigawatt of power. Furthermore, his analysis essentially rules out natural processes such as ball lightning. Despite the exhaustive efforts of Dr.

---

[45] https://frederikuldall.dk/wp-content/uploads/2024/02/jse_13_2_maccabee.pdf

Maccabee in his scientific analysis, he was subjected to the second problem which has traditionally plagued the science of ufology, that is the rejection by mainstream science who act as "gatekeepers" as they cherry-pick what is published and thus regarded as "actual" science. By keeping the vast majority of anomalous research out of the academic environment, it is thus treated as a pariah and oftentimes labeled as "pseudoscience."

In this particular instance, Dr. Maccabee, who has a Ph.D. in physics, was outright rejected when he first attempted to have his research published in *Applied Optics* in 1984 by an editor who appeared to be biased against the subject matter in its entirety and based his objection upon proclamation rather than objective evidence. Dr. Maccabee was not even afforded the opportunity to defend his position or appeal the rejection despite the fact that he constructed a comprehensive rebuttal of the editor's neglect of the scientific facts. A relative reluctance to consider anything out of the "ordinary" runs rampant throughout the scientific community, generally speaking. Few mainstream or what would be considered "legacy" scientists give any credence to exotic explanations of anomalous phenomena. Another example of this from Dr. Maccabee's work is his response to the 1997 Society for Scientific Exploration's review panel of anomalous radar and radar-visual hits. During this workshop which occurred in October of 1997, numerous high-quality cases were presented which involved anomalous radar hits, including a proportion of which were corroborated with visual sightings.

Despite being presented with a plethora of evidence, the members of the panel immediately vacated the possibility of extraterrestrial origin. It is almost as if the panel members would deny this possibility to the umpteenth degree despite evidence to the contrary. The ETs would literally need to land on the front lawn of the white house for them to consider this possibility! The panel, in this case, chalked up the radar and radar-visual targets to "atmospheric effects" but provided no valid evidence to support this claim. In his 38 page rebuttal, Dr. Maccabee meticulously provides detailed evidence to support the conclusion that atmospheric effects cannot account for the radar and radar-visual data. Some of the most compelling data included occurred on March 8th, 1995, at a military radar station in Switzerland, which included multiple radar targets traveling near mach three, all of which were descending almost vertically (elevation angle 2.90°) from an altitude of 21.7 km down to 6.2 km. At that time period, mach two was the limit for high-performance aircraft within the European operating theater. Dr. Maccabee concludes that some of these radar returns do indeed show properties consistent with intelligent control beyond that which is within the capabilities of human-made technology or nature.

The BLT research group is a non-profit established in 1999 and composed of several hundred researchers whose goal is the scientific analysis of physical changes which occur in soil and plants both within and near crop formations. Modes of information dissemination include peer-reviewed journals, lectures, articles, and the web. A sizable amount of their work deals with the analysis of soil samples from within crop formations, which overwhelmingly and disproportionately (in comparison to soil samples from outside the formations) contain large numbers of 10-40 micron diameter spheres that are composed of iron. Occasionally, larger particles of 40-50 microns in diameter are also found which display magnetic properties. Interestingly, all of these sphericles are spread out quite evenly along the perimeter of the crop circle formations and thus demonstrate soil deposition by a centrifugal force which results in a linear deposition (quantity of magnetic spherules increases from the center of the crop circle to the periphery of the crop circle).

Interestingly, Loeb et al. (2023) described spherules which were recovered from the seafloor in June of 2023, which have been determined to have extra-solar properties, including unconventional isotopes of iron and excessive amounts (approximately three orders of magnitude) of Beryllium, Lanthanum, and Uranium which are not seen within our solar system. The authors concluded that the samples studied might originate from a highly evolved magma ocean on a planet outside of our solar system. Another possibility the authors allude to is that of origination from supernovae and/or neutron star collisions. Lastly, the authors alluded to the possibility of an extraterrestrial technological genesis.

Fleming (2006) described anomalous objects which were detected during the space shuttle mission, STS-48 which occurred on September 15th, 1991. During a video recording of this mission which involved the space shuttle discovery, several objects were recorded which displayed abrupt movements and defied conventional physics in some respects. When this was brought to the attention of NASA, they basically gave the explanation that anomalies were the result of particles being pushed around by the force of the shuttle's thruster engines. Researcher Mark Carlotto later constructed time-exposure images from the mission which demonstrated that the trajectories of these "particles" indeed showed a curvature of radius, thus indicating lengthy periods of acceleration that cannot be explained by the brief bursts of energy that arise from the shuttles rocket thrusters exhibiting forces upon these particulate clusters. Based upon this, researchers outside of NASA concluded that the objects recorded during this space shuttle mission were more than likely self-powered, thus leaving open the possibility of artificial and/or exotic origin of these objects.

Although J. Allen. Hynek receives mixed reviews within the Ufologist community for his oftentimes overly skeptical view of cases which are regarded by most as truly anomalous; he must be credited with the current classification system, which is used by the majority of modern UFO research organizations and investigators. These include close encounters of the first kind, second kind, third kind, and fourth kind. These are given the acronym CE1, CE2, CE3, & CE4 accordingly. In CE1 cases, the flying object typically comes within 500 feet of the witness such that the witness is able to see the structural details of the craft. In these cases, there are no effects on the environment or the witness. Some researchers have noted that CE1 cases typically can provide some visual evidence about the craft's origin. Researcher Michael Schratt has suggested that the presence of structural features such as bolts, rivets, fasteners, pipes and other hardware are typical of reverse-engineered craft (top secret craft), whereas crafts that have a smooth, streamlined surface appearance are typical of ET origin. CE2 cases involve direct effects on the environment, humans, and animals. This includes effects from radiation, heat, and electromagnetic interference. The electromagnetic properties of UFOs commonly interfere with radios, vehicle ignitions, cell phones, and facilities which house nuclear missiles in preparation for launch. CE2 cases have the potential to cause injury to humans, plants, and animals.

An example of a classic CE2 case is that of prospector Stefan Michalak on the afternoon of May 20th, 1967, near Falcon Lake in Manitoba, Canada. He was performing his normal prospecting activities, collecting samples from a quartz formation in the area, something that he did regularly when he noticed two saucer-shaped crafts approximately 50 meters away. To his surprise, one of the craft made contact with the ground while the other flew away at a fast pace. Curiously, Mr. Michalak approached the craft, thinking at the time that this was a top-secret military craft. He noticed a hatch-like panel on the craft with a grid-like pattern and decided to inspect this further. Upon approaching this area of the craft, suddenly, the entire craft tilted on an angle towards him and he felt tremendous pain over his chest and abdominal region. He became ill later that night, and the weeks followed with symptoms of radiation sickness, including nausea, vomiting, diarrhea, and severe headache. He subsequently sought medical attention and was eventually treated at Mayo Clinic. However, the incident was reported to both Canadian and American officials, who were not able to offer a conventional explanation for the incident. Subsequent collections of metal fragments from the site of the landing were tested and found to have high levels of radioactivity. Unfortunately, Mr. Michalak dealt with chronic pain and dermatological sequelae for many years after the incident.

Close encounters of the third kind or CE3 cases are classified by the sighting of an actual alien or non-human being. Classic examples of these cases include the Roswell incident, the Aztec, New Mexico incident, and the Ariel case. In all three of these cases, witnesses saw a living or deceased being. Also included with the CE3 designation is the fact that sometimes there is an effect upon the witness, such as telepathic communication with the being, paralysis or a trance-like state. In fact, some of the children involved in the Ariel case described being in a trance-like state when the aliens were looking at them. They also described being communicated with by these aliens and given messages which took the context of the importance of taking care of the earth's environment and the importance of eliminating the polluting of our planet. Close encounters of the fourth kind, although not originally described by Dr. Hynek, but added on later when ufologists needed a way to classify abduction cases such as the Betty and Barney Hill case.

Known as CE4 cases, the key to these is that witnesses are taken against their will by non-human beings and are subjected to examinations, implantations, or targeted communications. For example, in the Betty and Barney Hill case, Betty described (under hypnosis) experiencing quite a spirited conversation with one of the beings on the craft. During this conversation, the being seemed to have a sort of "attitude" with her which was exemplified by the fact that when Betty asked the being where he was from, he responded by saying to betty something along the lines of "What good does it do for me to tell you where I am from if you don't even know where you are from!" With this statement, the being then tried to explain to Betty where exactly we are in the universe and more specifically, where we are within our little rung of the Milky Way Galaxy. In doing so, the being was able to communicate a star map to Betty which showed all of the common "trade routes" which are frequented by all of the different species of beings within our galaxy. Thus, Betty was given information about our galaxy during her and her husband's abduction which is the epitome of a CE4 encounter.

The Center for UFO Studies additionally describes three standard categories of UFO sightings which include nocturnal lights (NL), daylight discs (DD), and radar-visual sightings (RV). Nocturnal lights comprise the largest number of UFO sightings and typically involve orange, white, and red lights. They are typically circumscribed and void of any structural features. Additionally, we can say that cases which are agreed upon by researchers as genuine often involve maneuvers which are extremely fast (often upwards of mach three or mach four, 2300-3100 miles per hour) and involve hairpin and/or sharp, right-angle turns which would not be compatible with survival for human pilots. Hence, this brings the possibility of whether or not these vehicles are unmanned or possibly manned by artificial intelligence.

Daylight Discs or DD classification is traditionally that of a typical saucer-style craft with tapered and rounded features and minimal surface features. Typical flight patterns involve omicron or delta configurations. The former configuration is a low-power mode in which the underbelly of the craft is pointed toward the ground. This is the typical "hovering" mode of a flying saucer as it surveys the landscape. The latter configuration is the high-power mode used for interstellar travel whereby the underbelly of the craft is positioned on its side, allowing for the gravity emitters to interact with and manipulate space-time. This presumably allows for non-linear travel which is not dependent upon the speed of the craft, but instead folds the space-time fabric such that points A and B merge together. The final classification in the Hynek system is radar-visual sightings (RV). This is the corroboration of radar targets with the actual visual identification of airborne craft by one or more witnesses.

Typically, this involves sightings by civilian or military pilots. For example, on March 27, 1966, FAA and military radar operators at Muscogee County Airport and Fort Benning, respectively, witnessed a radar-visual sighting of an anomalous craft hovering over Columbus, Georgia. The object was described as changing its shape from a cigar shape to a triangular shape and displaying greenish-white lights. The expanded use of forward-looking infrared (FLIR) systems has significantly broadened the ability of military aviators to detect anomalous craft within their operating theater over the past several years, as we have seen with the Pentagon's release of the tic tac Gimbal and go fast videos from 2004, 2014, and 2015 respectively. These FLIR systems are able to construct an image on a camera based on the detection of thermal radiation.

Published in 1955, Project Blue Book special report #14 (PBB #14) is still regarded as the largest and most scientific undertaking of UFO data in the history of Ufology. To understand the importance of this study, one must become familiar with the tenants of performing credible research, which includes not withholding information that may support or refute your conclusions, upholding high ethical standards, and presenting statistical data as is, without manipulation, which would give artificial support to mundane data or suppressing the deleterious effects of confounding variables. Unfortunately, for the researchers who conducted this study, these standards were not upheld. Specifically, in their conclusions of the study, key statistics were withheld which refuted their official conclusion which basically stated that UFOs do not present a national security risk, nor do they possess evidence of advanced technology. Finally, they conclude that based on their analysis of available data, no evidence was found to support the hypothesis that UFOs are of extraterrestrial origin.

To the contrary however, when one digs into the data, it runs totally contradictory to said conclusions. Of the 4000 sitings initially enrolled in the study, a meticulous vetting process was separated into three refining steps in the initial data analysis which deducted a fair amount of nebulous sightings (800 to be exact). The remaining 3200 reports, when subjected to further vetting, which only includes actual sightings of true physical objects, reduces the final data set to 2199 sightings (Maccabee, 1979). Cases in PBB #14, which were statistically determined to have both good and excellent (unknown) object sighting reliability, were 24% and 33%, respectively.

Subsequent Chi-square analysis comparing the characteristics of knowns versus unknowns in the PBB #14 data set (characteristics included: color of the object, number of objects observed, shape of object, speed of object, duration of the observation, and brightness of the object) indicated significantly high values which would indicate that the chances of equivalence between knowns and unknowns is less than 1% among five of the six characteristics (brightness of the object was the only characteristic which did not demonstrate significance). Thus, results indicate that "unknowns" are truly independent of the "knowns" according to the Chi-square analysis.

Despite the significance of these results, PBB #14 researchers concluded that the majority of these "unknowns" were most likely "known" objects such as conventional aircraft, birds, natural phenomena, etc.... which could be identified if higher quality observations can be garnered and/or confounding variables can be eliminated. Despite the gravity of these results, the Air Force withheld this in the official press release of PBB #14. The importance of PBB #14 is therefore not reliant on its published conclusions because of their inherent flaws and omissions. Its importance lies in its data quantity, quality, and vetting process which accomplished the identification and isolation of high-quality sightings for subsequent scientific analysis and critique which has highlighted its inherent yet underreported value.

For perspective, although the most recent UAP report for the fiscal year 2024 written by members of the All-Domain Anomaly Resolution Office (AARO) was highly anticipated, the body of the report is quite anemic in comparison to PBB #14. The latter is a comprehensive analysis that includes 312 pages, chock-full of graphs, charts, statistical analysis, case details, object details, and a vetting process. In the AARO report, by contrast, details about the vetting process, methods of inquiry, and statistical analysis were scant and void of definition with regard to the 118 resolved cases during the reporting period of May 1st, 2023, through June 1st, 2024.

AARO reports that all of these cases were attributable to commonplace objects such as airplanes, satellites, balloons, birds, or drones. Moreover, a salient point is made within the body of the executive summary, which states that no evidence of aliens, alien technology, or alien activity has been found during their investigation. Since this statement is placed promptly into the report, it comes across to the reader as a preconceived conclusion. Hence, one must speculate whether AARO's analysis of cases is solely intended to fit this narrative. If their intentions were truly scientific, an offering of this conclusion would not be highlighted. If the intentions of the agency are truly objective and unbiased, no such mention of extraterrestrials would be included. We should expect our scientists to be open to any and all possibilities; this is the true spirit of science.

In November and December of 2014, reports of "drones" began to surface and were broadcast by major media outlets including ABC, CBS, NBC, FOX, CNN, News Nation, etc.... These were being reported by hundreds of citizens in the New Jersey and New York City metropolitan area initially, but over the course of several days, sightings began to spread into Maryland, Pennsylvania, Washington DC, Indiana, Illinois, Minnesota and this author anticipates this will continue to occur across the country. This event is occurring as this book is being written, and thus far, the authorities have offered no explanation. These objects have been described as very large, the size of a car or small school bus, fixed wing, and have been described as both silent and some which make noise. They have multiple lights flashing and are flying low, in some cases less than 100 feet above the ground in extremely densely populated areas in these metropolitan areas. Several theories have been floated, including the possibility that these are Russian or Chinese military drones (a claim that our government officially denies). Others claim these to be non-human technology or some secret technology that our own government is aware of. This would presumably be similar to some of the black budget technology which is tested at our various government installations in the southwest (i.e., Area-51, Tonopah test range, White sands, Area-51, etc....). Still, others claim that they could be originating from private companies testing some of the newest artificial intelligence.

Thus far, the response to these incursions from federal government officials has been lukewarm at best. The FBI has denied knowing anything about the origin of these vehicles and White House officials have basically denied any possibility that these are foreign actors. Interestingly, despite outwardly admitting that the origin of these craft is unknown, in the same breath, they are stating that there is essentially no public safety risk, which is an obvious concern considering the fact that these objects are hovering and maneuvering over very densely populated areas for up to eight hours at a time. Furthermore, some reports and video recordings have demonstrated swarms of these objects in the sky simultaneously, up to a dozen at a time, sometimes flying in formation. They only appear during the evening hours thus far between the hours of dusk to dawn.

Additionally, these have been witnessed by hundreds of people, similar to the major UFO sightings of the past, including the Phoenix Lights and the Hudson Valley event. These latter two cases are also events by which craft flew over heavily populated areas. This author has yet to formulate an opinion on these most recent events due to the fact that the situation is still evolving. However, a couple of curious thoughts come to mind. I question the theory that these recent sightings are the result of testing by private companies or government contractors, given the fact that a tremendous amount of liability exists by testing over heavily populated areas. If one of the craft were to crash, the risk to the public would be extreme in terms of loss of life and property. The other fact that is at odds with this theory is that no flight plans have been reported by the FAA. In fact, there have been no reports from any Federal or State agencies indicating that private or government aerospace companies would begin flight testing in this manner.

Doing this type of testing in a major metropolitan area would be highly irresponsible of any aerospace company, especially in today's climate of widespread lawsuits and litigation which occur even for the most trivial of matters (e.g., recall the 1994 case of the customer who filed a lawsuit against a fast food restaurant because of spilled hot coffee...). Another thought is the possibility that what we are seeing is an obvious result of Non-Human-Intelligence (NHI) to construct craft which have now more aptly "blended in" with what we commonly view as normal within our airspace. Think of the fact that back in the 20th century, the majority of the early sightings were classic, saucer-shaped craft which flew and hovered silently and displayed the classic right angle turns and quick acceleration which for all intents and purposes allowed them to stick out like a sore thumb.

As time went by, we then saw an evolution of craft into various shapes and sizes, such as triangular craft, V-shaped craft, circular orbs, tic-tac shapes, and even cubes within spheres. Perhaps what we are seeing now is just the next evolution by NHI to blend their craft into what we as humans perceive as somewhat normal; even adding noise production mechanisms to the crafts that have traditionally been silent could be an attempt to assimilate the public to this new reality of their obvious existence. The fact remains that this new era of UAP/UFO sightings is growing, with incursions even forcing some airports (both civilian and military) to halt air traffic for a period of time due to the quantity of incursions which are putting our conventional aircraft in a flight safety risk due to the possibility of midair collisions.

To date, this has already occurred at both Stewart Airport in Orange County, NY and at Wright-Patterson Air Force Base near Fairborn, Ohio, during the month of December 2024. The level of frustration among both the public and local government officials with these November/December 2024 incursions has been increasing due to the lack of response by White House officials, FBI officials, FAA, and Homeland Security officials. Public sentiment is largely untrusting of the government, and this is reflected in the comment sections of multiple online news agencies and blogs. Moreover, it seems that similar incursions have been happening with increased frequency all over the world, including multiple military air bases in the UK, including RAF Lakenheath, Mildenhall, and Felwell. Interestingly, the majority of incursions are being labeled as "unmanned aerial systems"; however, it is unclear how these sources have reached the conclusion that these craft are not piloted. The emerging public acceptance of the UFO reality is a far cry from what we had witnessed even 5 years ago when most mainstream news broadcasts either did not cover UFO stories at all or when they did, we basically got the “giggle” factor, along with cued X-files music. The topic was treated in an unserious manner, and its researchers were largely regarded as “pseudoscientists.” Thus, if one were to summarize the current status report, it would be an obvious evolution of ufology with regard to its importance in terms of science, national security, and aerospace safety.

References:

1) https://www.bltresearch.com/magnetic.php
2) https://cufos.org/types-of-ufos/classification-systems/

3) The Department of Defense All-Domain Anomaly Resolution Office, Fiscal Year 2024 Consolidated Annual Report on Unidentified Anomalous Phenomena Information Cut Off: June 1, 2024

4) Fleming, L.D. (2006). Examination of the Trajectories of Anomalous Objects Imaged During the STS-48 Space Shuttle Mission. The Journal of UFO Studies, Vol. 9, pp. 71-98.

5) Hall, R. (1980). UFO "Fingerprints": A Fourth Dimension. The Journal of UFO Studies, Vol. II, pp. 13-16

6) Loeb, A. et al (2023). Discovery of Spherules of Likely Extrasolar Composition in the Pacific Ocean Site of the CNEOS 2014-01-08 (IM1) Bolide.

7) Lore, Gordon. (1969). Strange Effects From UFOs. (Washington: NICAP).

8) Maccabee, B. (1999). *Atmosphere or UFO? A Response to the 1997 SSE Review Panel* Report. Journal of Scientific Exploration, Vol. 13, No 3, pp. 421-459, 1999.

9) Maccabee, B. (1979). Scientific Investigation of Unidentified Flying Objects: Part 1. The Journal of UFO Studies, Vol. 1, pp. 70-92

10) Project Blue Book Special Report, No 14, Air Technical Intelligence Center (ATIC) Wright-Patterson AFB, Ohio (1955)

Chapter-10
The Missing

In 2018, a multipart mini-series titled "Manifest" aired, which was about a passenger jet that suddenly experienced severe storms and turbulence on a routine flight that lasted several seconds. All the plane's passengers and crew are uninjured; however, upon arriving at their destination, they are met by law enforcement and government officials. It seems that their plane went missing, and five years went by before the plane suddenly reappeared on the same flight pattern. As a result, all the passengers on the plane arrived in a world in which they had not aged, but the world moved on without them. As a result, friends and family members aged, and some died. The drama then mainly deals with their assimilation back into their families and into society with several twists and turns and newly discovered superpowers that some passengers now have. This may seem quite far-fetched, but upon watching this drama series, one may recall the case of Malaysia Airlines flight 370 (MH370), a Boeing 777 which essentially vanished without a trace back on March 8th, 2014, while flying from Kuala Lumpur International Airport in Malaysia to the Capital International Airport in Beijing China.

On board this flight were 227 passengers and 12 crew members. The missing flight led to a search effort which covered a large area from just west of Australia through to the central Asian continent. Until this day, this incident is regarded as the greatest modern day aviation mystery. Despite numerous efforts to find the plane hosted by both governments and private agencies, no definitive proof of the plane has been recovered. It is as though the plane literally disappeared into thin air. There have been rumors that the plane was sabotaged, accidentally shot down, and even rumors that one of the pilots purposely used the plane for his own personal suicide mission. However, to date, the evidence of any of this has been mediocre at best. Over a decade has passed since the plane disappeared and we are no closer now to discovering its whereabouts than we were when the incident first occurred. Pieces of debris have been recovered in the years since the incident along the coast of the African continent and various islands within the Indian Ocean; however, there has been speculation about whether or not these pieces genuinely originate from MH370. The lack of any definitive evidence has led to speculation about what possibly could have happened to this flight and numerous theories of possible paranormal explanations have been floated.

When one considers the thousands of documented cases of UFO abductions and the rich history of missing planes and ships in the area of the Bermuda Triangle, it is not a far stretch to consider this as a viable possibility in the case of MH370, especially considering the lack of definitive evidence and paucity of progress regarding the location of this aircraft despite recovery efforts which have spanned thousands of miles and have cost tens of millions of dollars. It has become the most costly search for a missing aircraft in aviation history. Researchers of anomalous phenomena have proposed theories about the possibility that the aircraft slipped into a wormhole or a portal of some sort and will eventually reappear at some point in the future intact.

Similar stories have floated through the anomalous lore and include the legend of Pan American Flight 914. This was a DC-4 aircraft with 61 passengers plus crew members on board which supposedly traveled from New York to Miami, Florida, on July 2nd, 1955. The flight was described as normal and the weather was clear. Generally, a good day for flying. Midway through the flight, air traffic control (ATC) lost communication with the pilot and the aircraft disappeared off radar screens. The flight was supposed to take three hours from New York to Miami, Florida, but at the time it was supposed to land, ATC at the Miami airport did not detect the flight on their radar, nor did they establish any communications with the aircraft. Again, just like MH370, this flight apparently vanished into thin air. An extensive investigation took place on land and in the Atlantic Ocean, but no signs of the craft or of its passengers were ever found. Thus, flight 914 was presumed to have crashed, and eventually, the official statement was just that, along with the presumed death of all of the passengers and crew on board.

37 years later, on September 9th, 1992, ATC in Caracas, the capital city of Venezuela, witnessed a target pop up on the radar out of nowhere. Confused about this, the controller checked his instrumentation and verified that his data was correct. After about 10 minutes of tracking the target on radar, visual confirmation of a DC-4 aircraft on approach was seen by multiple ATC personnel. The controllers were indeed perplexed by this visual because the aircraft approaching was an older-style DC-4 with propeller engines instead of modern turbine engines. The old propeller engines on this size aircraft were almost obsolete in 1992 and thus rarely seen. To complicate matters, the pilot of the approaching aircraft's first attempt at contact with ATC commenced with the question, "Where are we?" At this point, the ATC supervisor, Juan de la Corte, took charge of the situation and replied to the pilot requesting identification of the aircraft, along with the place of origin and destination. To the shock of the ATC supervisor, the pilot responded by confirming their flight was Pan American flight 914, destination Miami, Florida, status post departure from New York City, on July 2nd, 1955. The pilot also confirmed a total of 61 souls on board, including passengers plus crew. The ATC supervisor was very concerned at this point, as he suspected that the pilot was having some sort of mental breakdown due to his obvious confusion regarding the date and current location. In light of these concerns, the ATC supervisor thought it best to not push this issue while the aircraft was still

in flight and on the final approach since this is a critical time of an aircraft's flight plan. The safety of the passengers and crew were the top priority of ATC, thus maintaining professionalism was salient in this goal.

Once the DC-4 safely landed, it was directed to taxi to a gate assignment for passengers and crew to deplane. Noticing quite clearly that other aircraft upon the tarmac were clearly different from the typical propeller-style crafts of the 1950s, the captain of flight 914 began inquiring, and the ATC supervisor, at this point, disclosed the current date of September 9th, 1992. Upon hearing this, the pilot requested clarification from ATC who reiterated the current date and time. When he heard this, he momentarily cut communication with the tower for several seconds. The next communication from the captain pilot was a demand for the approaching ground crew and security to stay away from the aircraft. Then, suddenly, the pilots began taxiing back out towards the runway and declared over the radio that they were leaving. Despite unanswered pleas from ATC, the plane departed without clearance and quickly disappeared on radar. There has been much dispute about the validity of this story over the years. Records searches have revealed a paucity of documentation regarding the existence of this particular flight. However, stories such as this rarely are produced out of thin air. Typically, there are bits and pieces of truth to these tales. The disappearance of MH370 has indeed brought some renewed intrigue into the Pan American 914 case however.

The Bermuda Triangle has been a popular subject because of the many ships, planes, and people that have gone missing in this region and are thought to have become victims of an unworldly origin. There has been speculation about alien abductions, parallel universes, wormholes, and other dimensions. Geographically, the Bermuda Triangle includes an area with an estimate of between 500,000 and 1,500,000 square miles outlined by fictitious lines which run from the island of Bermuda to the southeast coast of Florida to the northern tip of Puerto Rico. It includes areas of the Bahamas and the Caribbean, and there is speculation that it includes the lost city of Atlantis. As far back as 1492, Christopher Columbus reported anomalies in compass readings and sightings of strange lights when he and his crew of sailors were passing through the Sargasso Sea, part of which lies within the triangle.

The name "Bermuda Triangle" was coined by the author Vincent H. Gaddis which he wrote for a 1964 article in Argosy magazine. In that article, Gaddis documented multiple aircraft and sea vessels which have vanished while traveling through this area of the world dating back to the mid-nineteenth century. Theories about strange clouds that cause microbursts of energy, seas of bubbling methane, and underwater alien bases have been cited by various researchers and enthusiasts as the cause for these missing planes and ships. However, to date, there has been no consensus. The USS Cyclops, the biggest ship in the Navy, went missing in March of 1918 and vanished into thin air in March of 2018, its wreckage never to be found over 100 years later (see figures 1 & 2). Many theories have been presented regarding the ship's disappearance, which include sabotage by unruly crew members, or an unstable captain who sunk the ship into the trench of Puerto Rico, an attack from the German military U-boats or mines[46], or swallowed by a giant octopus. The few facts which are eerie about this disappearance are the absence of any distress calls from the ship and the unresponsiveness of the Cyclops despite the many messages that were sent to the ship from multiple ships in the immediate vicinity. Thus, despite logical explanations for the loss of the ship which have been proposed, the fact remains that it appears to have simply vanished without a trace of evidence to support any of the "logical" explanations of its disappearance. It is for this reason, we should not rule out a novel causality for the disappearance of the USS Cyclops.

[46] The theory of an attack by the German military has been ruled out since there has never been any evidence that German vessels or weapons were ever located in this vicinity

Figure-1. USS Cyclops. Source: https://www.loc.gov/resource/hec.00089/(Library of Congress).

Register and Tribune's Free War Time Cooking School, April 22 to 26

The Des Moines Register

(Formerly The Register and Leader)

300 MISSING ON NAVAL COLLIER

NEW DRAFT ORDER WILL TAKE 1,910 MEN FROM STATE

British Holding Firmly in Most Critical Sector

GERMANS CONTINUE ATTEMPT TO FORCE AMERICAN TROOPS

"PAYING TOO MUCH FOR HIS WHISTLE"

SEVEN IOWANS ON CYCLOPS OVERDUE SINCE MARCH 13

Figure-2. Newspaper article from the Des Moines Register, April 15th, 1918, about the disappearance of the USS Cyclops. Source:

https://medium.com/iowa-history/unsolved-mystery-uss-cyclops-vanished-100-ye ars-ago-96243368df03

The 1977 film Close Encounters of the Third Kind depicts a scene towards the end of the film in which the "mother ship" brings home multiple individuals who have gone missing, including soldiers, sailors, pilots, and civilians. All of the individuals are intact, uninjured, and have not aged beyond the age they were at the time of their abduction. The next case may depict such an instance whereby the abduction of multiple individuals by some anomalous force may have occurred. In December of 1880, the Ellen Austin passenger ship set sail from Liverpool, England, to New York, bringing emigrants from Europe to the new world of the Americas. Along this journey, the ship encountered forceful winds which forced the captain (Captain Griffin) to bring the ship southward a bit along the journey to make up time. In doing so, the ship entered the northern portion of the Sargasso Sea. This region is adjacent to and includes a portion of the Bermuda Triangle. Approximately midway through his typical 30-day journey, Ellen Austin's crew spotted another ship (a schooner-type ship, see Figure 3) floating aimlessly in the distance. Upon approach, the crew of the Ellen Austin signaled to the lone ship but did not receive a response, and they never saw any of the crew members upon the ship. After two days of trying to establish communications, the captain of the Ellen Austin ordered a small team of his crew to go aboard the schooner to inspect it. Upon boarding the ship, crew members found everything to be intact which included plentiful supplies and a cargo of timber. However, not a living soul was aboard the ship. There was no sign of any struggle or conflict about the schooner. The crew had simply vanished without a trace. Upon discovery of a valuable payload, the captain of the Austin orders his crew to remain upon the abandoned ship as they escort it to the mainland of the United States. After some time, the two vessels were separated due to a combination of fog and rough seas.

Upon reconnecting with each other, the captain and crew of the Ellen Austin once again approached the other ship, expecting to find their crew members aboard. However, to their shock, the schooner was totally crewless again, with no sign of the Ellen Austin crew members who were instructed to remain on the ship for the duration of the voyage. The captain of the Austin now frustrated, then ordered a second crew of very reluctant crew members to go aboard the schooner to resume the duties of guarding its cargo and escorting it to the mainland. As a precaution, both crews on each ship were instructed to signal to each other by ringing a loud bell at regular intervals, to communicate and provide reassurance that everything was running smoothly. At first, this communication proceeded regularly, however after several hours, a thick fog set in and the bell ringing from the schooner abruptly stopped. The crew of the Ellen Austin continued to ring their bell throughout the night but never received a response. After a while, they could also no longer see the lantern aboard the schooner ship. When dawn finally arrived, the fog lifted and the schooner, along with the second crew, had vanished and was never to be found again. The paucity of official record keeping during this period of time calls many of the details of this case into question. However, it remains steadfast as part of the lore surrounding the Bermuda Triangle.

One of the most compelling cases involving the Bermuda Triangle is the flight-19 incident. On December 5th, 1945, a group of five TBM Avenger torpedo bomber aircraft with a total of 14 crew members on board mysteriously went missing while taking part in a routine training flight which departed from the Naval Air Station at Fort Lauderdale, Florida under the command of squadron leader, Lieutenant Charles C. Taylor. Lieutenant Taylor was an experienced airman with over 2500 hours in the cockpit, which also included multiple tours of combat during World War 2 over the South Pacific Ocean. This particular training exercise basically followed a right triangle, initially following a due east course for 56 miles out over the Atlantic Ocean to engage in bombing exercise training in the region of Hen and Chickens coral reef. After the training exercise was complete, the five aircraft were then directed to fly another 67 miles east and then turn north for another 73 miles and fly over Grand Bahama Island.

The last leg of the trip required the squadron to turn and head southwest for 120 miles and return to Fort Lauderdale, Naval Air Station. As the squadron proceeded east after their bombing exercise, suddenly the navigation equipment in all five planes malfunctioned simultaneously and as a result, the planes no longer knew which direction they were headed. Their communications with air traffic control at NAS Fort Lauderdale became increasingly filled with static, leading ATC to believe the squadron was headed further out to sea in an eastward direction. By this time, the squadron leader was so confused that, at one point, he thought he was in the Gulf of Mexico. Interestingly, ground-based communications had increasing difficulty communicating with flight-19; however, they were able to monitor the conversations between the pilots, in each of the five planes. Communications between the pilots grew increasingly concerning as they were making statements to each other noting distorted appearances of the ocean, the sun, and the sky. Furthermore, none of the pilots were able to make a visual identification of the landmarks below. This concern was particularly concerning with regard to the team commander, Lt. Taylor, who had the most flight time hours, 2500, compared to the remainder of the pilots.

A rescue aircraft, the Martin Mariner, with 13 crew members aboard, was sent to the area where the squadron was thought to have disappeared (somewhere along the due east leg of the mission). Shortly after take off, while heading out to sea, the Lieutenant commander of the Mariner made radio contact with ATC to notify them of strong winds above 6,000 feet altitude. That was the final transmission from the Mariner which suffered the same fate as the Avengers. There was never any shred of evidence found of any of the aircraft that went missing that day, despite an enormous search and rescue mission covering 380,000 square miles of sea and land by hundreds of ships, boats, submarines, private and military vessels, and 307 aircraft. This was one of the largest search and rescue missions in history and it resulted in nothing. Not even a shred of ancillary evidence, such as an oil slick, was ever found despite a total of 4100 hours spent searching over several weeks. The mysterious nature of this disappearance of six aircraft in the same vicinity without any evidence of a crash is what led investigator and author Charles Berlitz to speculate about a possible supernatural cause of this disappearance in his 1974 book, *The Bermuda Triangle, an incredible saga of unexplained* disappearances.

The mystery of this case was further propagated by the appearance of the five Avenger aircraft again in the film *Close Encounters of the Third Kind,* which supposedly materialized many years later in 1977 in the Sonora Desert of Mexico intact and without any evidence of human remains. Towards the end of the film, all of the pilots are returned by the "Mother Ship" also intact and unaged. Although unsubstantiated, one must wonder if perhaps bits and pieces of truth are perhaps purposely infused into popular cinema. The seal of mystery was finalized when the Opa-Locka Naval Air Station in Miami received a faint voice message from an unknown aircraft at 7 pm, identifying themselves as "FT," "FT," and then quickly fading away and never heard from again. FT was the call letters for all five of the flight-19 Avengers. In fact, the squad leader, Lt. Taylor's plane was FT-28. This message was received two hours after each of these aircraft would have presumably run out of fuel. Many researchers have commented that this last eerie message could have been received from another dimension or perhaps from somewhere else within space-time. At this point, 80 years later and still no trace of these aircraft, one must consider the current context of science which has uncovered such possibilities as wormholes, multiverse theory, time travel, and non-human intelligence; topics which only were considered science-fiction during the majority of the 20thCentury.

Figure-3. Depiction of a Schooner ship. Source: https://www.loc.gov/item/2016806405/ (Library of Congress)

The case of Thomas Messick Sr., age 82, a hunter from upstate New York, is one of the most perplexing cases of a missing person in recent times. On November 15th, 2015, Thomas went hunting with a group of six longtime friends and family members in the Lake George area of upstate New York. Thomas was an experienced hunter and had been hunting in these woods his entire life. Furthermore, Thomas was a military veteran in the army where he served as a paratrooper in the 82nd airborne division. He was an experienced hunter who, for many years, taught survival training and hunting skills. The men he was hunting with that day he was friends with for 55 years and they had hunted numerous times together. On this particular day, they decided to hunt at Brant Lake. Thomas had a heart condition, so the friends decided Thomas would be safer to sit stationary while the remainder of the group went into the woods and drove the deer towards him. Thomas was positioned sitting on a tree stump and was only about 40 yards from the main road. After hunting for about two hours, the hunters all returned to their meeting spot, but Thomas was nowhere to be found. A massive search ensued which involved hundreds of search and rescue personnel, the FBI, and search dogs over several weeks searching the entire region, over 29 square miles. Despite this extensive air, land, and water search and rescue effort, not a shred of evidence was ever found of Thomas. Some officials noted there was no sign that Thomas had ever been in the forest that day; it was as if he had vanished into thin air.

Foul play was essentially ruled out due to the fact that there was no evidence of a struggle, and Thomas was armed with a rifle. Strangely, some of the search and rescue personnel admitted eerily on that day there was an absence of wildlife sounds in the forest. One search & rescue person, when interviewed, was quoted as saying, "It was almost like something silenced the woods." Moreover, some of the other hunters in the area that day reported hearing strange sounds in the forest on that day. When asked to describe the sounds, the hunters could not elaborate other than admitting it was a sound they had never heard before. Some thought perhaps the sound was coming from a higher elevation, possibly from the top of a hill. This remains an open case and leaves much room for speculation about what happened to Mr. Messick. The absence of wildlife sounds in the forest which was reported by search and rescue personnel, is quite mysterious, as this has also been frequently reported in UFO sightings that occur in similar settings.

This chapter on *The Missing* is certainly not complete without a word about ET contact and abduction cases. Although the details of abduction and ET contact cases indeed comprise volumes of evidence-based literature[47], we will make a few salient points here. One will hear common criticisms in popular culture and media which is directed towards

---

[47] For in-depth analysis, I would suggest any of the various works by Kathleen Marden such as Extra-Terrestrial Contact; what to do when you've been abducted (2019).

abductees (i.e., "Alien Abduction Experiencers") is the notion that they collectively suffer from mental illness. It is important for readers to know that this criticism is entirely false. Multiple studies have investigated this claim and have found no significant differences in diagnosed mental illness among AAEs when compared with the general population (Spanos et al, 1993; Parnell & Sprinkle, 1990; Rodeghier et al., 1991; Bartholomew et al., 1991, as cited in Kelly-Romano, 2006). Most of the recalled information from AAEs is divulged while under hypnosis (80% to 90%) and includes such details such as physical contact or examinations conducted by ET beings, methods of abduction, methods of communication between AAEs and ET beings, and messages for humanity given to AAEs. Research has demonstrated commonalities among AAEs, including the commencement of contact with ETs in childhood, a belief they have lived a past life, patterns of repeated encounters with ETs, a belief they hold some degree of psychic ability, more than one person within families with similar experiences, belief in some degree of self-healing abilities, and the belief that their contact with ETs has been largely a positive experience (Kelley-Romano, 2006).

The aforementioned cases of "the missing" are less than a fraction of the thousands of mysterious and unexplained occurrences of objects and people that vanish from the face of the earth without a trace. Our inner sense of logic begs us to find a legitimate explanation; however, in many instances, commonplace explanations do not fit. A quote from author and researcher Charles Berlitz summarizes these mysteries comprehensively; "We cannot retreat from the search for knowledge or new explanations, either in this world or beyond." Hence, it should be the goal of all scientists and researchers to keep an open mind and consider *all* possibilities.

References:

https://apeejay.news/mystery-of-the-bermuda-triangle-ellen-austin/

Bartholomew, Robert E, Keith Basterfield, & George S Howard (1991) UFO abductees and contactees; Psychopathology or fantasy proneness? Professional Psychology, Research and Practice, 22 (3). Pp 215-222.

Berlitz, C. (1974). The Bermuda Triangle, an incredible saga of unexplained disappearances Doubleday & Company, INC, Garden City, New York.

Dennis, C. (2022). Bermuda Triangle, A True Story of Strange Events From the Bermuda Triangle. Regina Loviusher.

https://www.history.com/news/bermuda-triangle-uss-cyclops-mystery-world-war-i

Kelley-Romano, Stephanie (2006). A Report on the Demographics and Beliefs of Alien Abduction Experiencers. Journal of UFO Studies, (9) pp: 1-20

https://www.outdoornews.com/2023/11/15/eight-years-later-search-for-2015-missing-adirondack - hunter-continues/

Parnell, June & Leo Sprinkle (1990) Personality Characteristics of persons who claim UFO experiences. Journal of UFO Studies, (2) pp: 45-58.

Rodeghier Mark, Jeff Goodpaste, & Sandra Blatterbauer (1991) Psychosocial characteristics of abductees; Results from the CUFOS abduction project. Journal of UFO Studies. 3, pp: 59-90

Spanos, Nicholas P, Patricia A Cross, Kirby Dickson, & Susan C Dubreurl (1993) Close encounters; An examination of UFO experiences. Journal of Abnormal Psychology. 102, pp: 624-632

Made in the USA
Monee, IL
03 November 2026

67378724R00105